Introduction to Probability

James E. Huneycutt, Jr.
North Carolina State University

CHARLES E. MERRILL PUBLISHING COMPANY
A Bell & Howell Company
Columbus, Ohio

MERRILL MATHEMATICS SERIES

Erwin Kleinfeld, *Editor*

Published by
Charles E. Merrill Publishing Company
A Bell & Howell Company
Columbus, Ohio 43216

ISBN: 0-675-08960-3

Library of Congress Catalog Card Number: 72-97005

AMS Classification Number: 6001

1 2 3 4 5 6 7 8 9 — 79 78 77 76 75 74 73

Printed in the United States of America

Preface

This book is intended as a text for a one-semester junior-senior course in probability carrying a prerequisite of a calculus course which includes multiple integration and power series (although these topics can be taken concurrently). It was developed while teaching such a course for several semesters to mathematics and engineering students at North Carolina State University.

The book is devoted more to presenting a mathematical structure in which problems can be solved than to deriving a large collection of formulae for solving particular types of problems. The text is incomplete in the sense that not all statements are proved; however, I have tried to indicate just what holes are left uncovered so that they could be filled in by a student with a more extensive background. Throughout, I have tried to present probability in an elementary fashion, but in a way that would not need *correcting* later, if the student takes a more advanced course.

The problems at the end of each section are purposefully not presented in order of difficulty or in an order reflecting the organization of the text material in that section. Those exercises which will be needed later in the text material are indicated by an asterisk (*).

Special Symbols and Abbreviations

$\mathscr{B}(\mathsf{R})$	28
$C(n, m)$	39
Cov (X, Y)	172
Δ	5
$\partial(A)$	143
$f^{-1}(C)$	71
N	2
p.m.s.	17
$\mathscr{P}(X)$	11
$P(n, m)$	36
$P(A \mid E)$	48
Q	2
R	2
r.v.	70
$\rho_{X,Y}$	176
σ-algebra	12
σ-ring	11
$\sigma^2{}_X$	126
σ_X	126
χ_A	74
Z	2
$\cup$	3, 8
$\cap$	3, 8
$\setminus$	4
A^c	5
$\Rightarrow, \Leftrightarrow$	4
$\times$	33
$!$	35

Contents

1 Sets   1

    1.1   Preliminary Notions   1
    1.2   Collections of Sets   6

2 Probability Measure Spaces   16

    2.1   Basic Definitions   16
    2.2   Discrete Probability Measure Spaces   22
    2.3   Other Probability Measure Spaces   27

3 Counting   33

4 Conditional Probability and Independence   48

    4.1   Conditional Probability   48
    4.2   Independence   55
    4.3   Sequences of Independent Events   62

5 Random Variables   69

    5.1   Introduction and Definition   69
    5.2   Distribution Functions   75
    5.3   Classification of Random Variables   81
    5.4   Diversion into Analysis   99

6 Expectation   106

    6.1   Definitions   106

6.2   Expectation of a Function with Respect to a
        Random Variable   112
6.3   Moments of a Random Variable   120

7   Jointly Distributed Random Variables   131

7.1   Joint Distribution Function   131
7.2   Classification of Jointly Distributed Random Variables   137
7.3   Independent Random Variables   151
7.4   Functions of Several Random Variables   159

I

Sets

1.1 Preliminary Notions

Probability is concerned with the assignment of numbers to possible occurrences in accordance with the likelihood of the occurrence. Since this is a rather vague concept, we may get perhaps a better understanding of the process of assignment if we abstract the salient aspects of this process into a mathematical structure. The occurrence mentioned above will correspond to a set; the process of assignment will be dealt with in chapter 2.

A *set* may be thought of as a collection of objects—a collection so well defined that given any object, it can be determined whether or not the object is in that collection. Sets are usually denoted by upper case letters (such as A, B, C) and the objects or *elements* of sets are usually denoted by lower case letters (such as a, b, c). To denote that an object a is an element of a set A, we shall write $a \in A$; if a is not an element of A, we shall write $a \notin A$. A set A is said to be a *subset* of a set B (denoted $A \subseteq B$) provided that each element of A is also an element of B. We shall say that a set A is equal to a set B ($A = B$) provided that $A \subseteq B$ and $B \subseteq A$. This simply says that A and B are two symbols denoting the same set.

A set may be defined by merely listing all of its elements; for example $A = \{1, 2, 3, 4\}$ in which the braces ("{" and "}") are used simply to gather together the members of A. In many cases, however, an explicit listing is not possible. If N denotes the set of natural numbers (positive integers), we may write $N = \{1, 2, 3, \ldots, n, \ldots\}$. For other cases, even the indication of a list is not possible; if (a, b) denotes the set of all real numbers between the numbers a and b, then we may write

$$(a, b) = \{x \mid x \text{ is a real number and } a < x < b\}.$$

The vertical line is read "such that." Here we define the set by giving some rule that characterizes the elements of that set. As another example, let Z denote the set of all integers, then

$$Q = \{x \mid x = p/q \quad \text{where } p \in Z, q \in Z, \text{ and } q \neq 0\}$$

is the set of all rational numbers.

There are several sets which occur so often that it is desirable to have special symbols for them:

    N denotes the set of all positive integers,
    Z denotes the set of all integers,
    Q denotes the set of all rational numbers,
    R denotes the set of all real numbers.

If a and b are real numbers, then

$(a, b) = \{x \mid x \in R, a < x < b\}$ is called a bounded open interval,

$[a, b] = \{x \mid x \in R, a \leq x \leq b\}$ is called a bounded closed interval,

$(-\infty, a) = \{x \mid x \in R, x < a\}$ is called an open, left infinite interval,

$(-\infty, a] = \{x \mid x \in R, x \leq a\}$ is called a closed, left infinite interval.

The reader should be able to invent reasonable definitions for the half-open, half-closed intervals $(a, b]$ and $[a, b)$, as well as the infinite intervals (a, ∞) and $[a, \infty)$. We shall not define such symbols as $[-\infty, a)$, preferring to consider "∞" and "$-\infty$" as concepts rather than objects of a set.

If a is less than b in the preceding paragraph then each of the bounded intervals defined there contains an infinite number of elements. If b is less than a, however, we have defined a set which has *no* elements. The set which contains no elements is called the null or empty set and is denoted by $\varnothing$. There is some question among mathematicians as to whether a collection with nothing in it legitimately can be considered a collection. Suppose you are walking down the street and meet a friend who informs you that he is the proud possessor of a stamp collection. You congratulate him on having an interesting hobby and ask him how many stamps are in his collection. He answers, "None." Can you reasonably agree that he has a stamp collection?

Some mathematicians accept the existence of the empty set and some reject it. In this book we shall use the empty set if for no other reason than that it often allows us to consider questions about sets without investigating several special cases. If use of the empty set *as a set* violates your mathematical conscience, you may consider it as shorthand for a concept. For example "$\{x \in \mathsf{R} \mid 1 < x < 0\} = \varnothing$" means that there are no real numbers greater than 1 but less than 0.

If A and B are sets, then we define the *intersection* of A and B by

$$A \cap B = \{x \mid x \in A \text{ and } x \in B\}.$$

For x to be an element of $A \cap B$ it must be in both A and B. The intersection is illustrated as the crosshatched area in figure 1.1. We shall say that A and B are disjoint if they have no element in common, i.e., $A \cap B = \varnothing$.

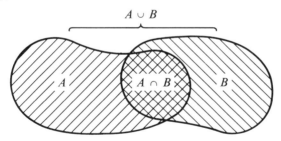

Figure 1.1

The *union* of the sets A and B is the set

$$A \cup B = \{x \mid x \in A \text{ or } x \in B \text{ (or both)}\}.$$

The following result gives two important relationships between $\cap$ and $\cup$.

DISTRIBUTIVE LAWS: If A, B, and C are sets, then

(1) $A \cap (B \cup C) = (A \cap B) \cup (A \cap C)$

(2) $A \cup (B \cap C) = (A \cup B) \cap (A \cup C)$.

Proof: We shall prove (1) and leave (2) as an exercise. If S_1 and S_2 are statements, then we shall abbreviate "S_1 implies S_2" or "if S_1, then S_2" by "$S_1 \Rightarrow S_2$"; we shall abbreviate the statement "S_1 is equivalent to S_2" by "$S_1 \Leftrightarrow S_2$". To show that (1) is true we must show that $A \cap (B \cup C)$ and $(A \cap B) \cup (A \cap C)$ contain exactly the same elements.

$$x \in A \cap (B \cup C) \Leftrightarrow x \in A \quad \text{and} \quad x \in B \cup C$$

$$\Leftrightarrow x \in A \quad \text{and} \quad (x \in B \text{ or } x \in C)$$

$$\Leftrightarrow (x \in A \text{ and } x \in B) \quad \text{or} \quad (x \in A \text{ and } x \in C)$$

$$\Leftrightarrow x \in (A \cap B) \quad \text{or} \quad x \in A \cap C$$

$$\Leftrightarrow x \in (A \cap B) \cup (A \cap C). \quad \square^1$$

For two sets A and B, we define the *complement of B in A* by

$$A \backslash B = \{x \mid x \in A \text{ but } x \notin B\}.$$

This set is sometimes called the *relative complement* of B in A and is illustrated by the hatched area in figure 1.2.

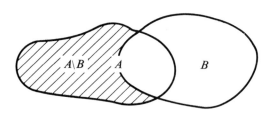

Figure 1.2

The operations of $\cap$, $\cup$, and $\backslash$ are related by the following result.

---

[1] We shall use the symbol "$\square$" to denote the end of a proof.

DEMORGAN'S LAWS: If A, B, and C are sets, then

(1) $A\backslash(B \cup C) = (A\backslash B) \cap (A\backslash C)$

(2) $A\backslash(B \cap C) = (A\backslash B) \cup (A\backslash C)$.

Proof: Again we shall leave (2) as an exercise and prove only (1).

$$x \in A\backslash(B \cup C) \Leftrightarrow x \in A \quad \text{but} \quad x \notin B \cup C$$
$$\Leftrightarrow x \in A \quad \text{and} \quad x \notin B \quad \text{and} \quad x \notin C$$
$$\Leftrightarrow x \in A \quad \text{and} \quad x \notin B \quad \text{and} \quad x \in A \quad \text{and} \quad x \notin C$$
$$\Leftrightarrow x \in A\backslash B \quad \text{and} \quad x \in A\backslash C$$
$$\Leftrightarrow x \in (A\backslash B) \cap (A\backslash C). \quad \square$$

The reader who has already been exposed to some of the elementary concepts of set theory may have noted the absence of any mention of the "universal set" in this text. This has been intentional, because very often it is not clear exactly what the universal set is. This makes it difficult to talk about such things as an "absolute complement" of a set. For example, the complement of the set of even integers could be the odd integers, all rationals except the even integers, or all reals except the even integers. Each change of the universal set leads to a change in the concept of complement.

If some set X has been specified and we are considering only subsets of X, then for $B \subseteq X$, we may denote $X\backslash B$ by B^c and call it the *(absolute)* *complement of B.* For example, in DeMorgan's Laws above, when $A = X$, we have

$$(B \cup C)^c = B^c \cap C^c$$
$$(B \cap C)^c = B^c \cup C^c.$$

Exercises:

1. Convince yourself that each of the following is true. Let A, B, and C be sets.
   (a) $\varnothing \subseteq A$
   (b) If $A \subseteq B$ and $B \subseteq C$, then $A \subseteq C$
   (c) $A \cap (B \cap C) = (A \cap B) \cap C$
   $$\text{and} \quad A \cup (B \cup C) = (A \cup B) \cup C$$
   (d) $A \cap B = B \cap A$ and $A \cup B = B \cup A$
   (e) $A \cap A = A$ and $A \cup A = A$

(f) $A \cap \varnothing = \varnothing$ and $A \cup \varnothing = A$
(g) $(A \cap B) \subseteq A$ and $A \subseteq (A \cup B)$
(h) $A\backslash A = \varnothing$, $A\backslash\varnothing = A$ and $\varnothing\backslash A = \varnothing$
(i) $(A\backslash B) \cap B = \varnothing$ and $A\backslash B = A\backslash(A \cap B)$

2. Prove (2) in the Distributive Laws.

3. Prove (2) in DeMorgan's Laws.

4. Let A and B be sets. Prove
    (a) $(A\backslash B) \cup (A \cap B) = A$
    *(b) $(A \cup B)\backslash(A \cap B) = (A\backslash B) \cup (B\backslash A)$. (This set is called the *symmetric difference* of A and B and is sometimes denoted by $A \triangle B$.)

5. If $A = \{n \in \mathsf{N} \mid n$ is divisible by 2$\}$ and
    $B = \{n \in \mathsf{N} \mid n$ is divisible by 3$\}$, then what set is $A \cap B$?

6. Let $A = (a, b]$ and $B = (c, d]$. What are the following sets? (Consider several relations among a, b, c, and d, such as $a < c \leq b < d$.)
    (a) $A \cap B$
    (b) $A \cup B$
    (c) $A\backslash B$

1.2 Collections of Sets

If we are interested only in the operations of addition and multiplication, then the set of positive integers may be a sufficient system of numbers in which to work. However, if we wish to consider subtraction or division, we may need a more complex system such as the set of rationals. Similarly, if we are interested only in the operations of union and intersection, a relatively simple class of sets may suffice; but for consideration of other set operations, we may need a more complex class of sets. In this section we consider various types of classes of sets—classes which are complex enough to allow us to perform certain types of set "arithmetic."

Definition 1:   Let X be a set and $\mathscr{L}$ be a non-empty collection of subsets of X. $\mathscr{L}$ is a *lattice* provided

(1) $\mathscr{L}$ is closed under $\cup$: $A \in \mathscr{L}$ and $B \in \mathscr{L} \Rightarrow A \cup B \in \mathscr{L}$,

(2) $\mathscr{L}$ is closed under $\cap$: $A \in \mathscr{L}$ and $B \in \mathscr{L} \Rightarrow A \cap B \in \mathscr{L}$.

Example: Let X be arbitrary and $\mathscr{L}$ be the collection of all finite subsets of X (A is finite if either $A = \varnothing$ or there is a one-to-one correspondence between A and $\{1, 2, \ldots, n\}$ for some $n \in \mathsf{N}$).

If $A \in \mathscr{L}$ and $B \in \mathscr{L}$, then A and B are finite subsets of X. Thus $A \cup B$ and $A \cap B$ are finite subsets of X, so $\mathscr{L}$ is a lattice.

Example: Let $X = \mathsf{R}$ and let $\mathscr{L} = \{(-\infty, a] \mid a \in \mathsf{R}\}$.

If $A \in \mathscr{L}$ and $B \in \mathscr{L}$, then $A = (-\infty, a]$ for some $a \in \mathsf{R}$ and $B = (-\infty, b]$ for some $b \in \mathsf{R}$. Now $A \cup B = (-\infty, a] \cup (-\infty, b] = (-\infty, c]$ where c is the larger of a and b; $A \cap B = (-\infty, a] \cap (-\infty, b] = (-\infty, d]$ where d is the smaller of a and b. Thus both $A \cup B$ and $A \cap B$ are members of $\mathscr{L}$.

We must stress that a lattice of subsets of X contains *subsets* of X, and not *elements* of X. There is a distinction between a subset and the elements of that subset—a subset is an entity itself. This was recognized by an ancient Chinese philosopher who observed that if one has a brown cow and a black horse, then he has three things; for the cow is one, the horse another, and the two together still a third. In the preceding example the set $(-\infty, 2]$ is a member of $\mathscr{L}$ but 1 is not a member of $\mathscr{L}$ even though $1 \in (-\infty, 2]$. $\mathscr{L}$ consists of sets, not numbers. In the same example, the set $[0, 1]$ is a subset of $(-\infty, 2]$ which is a member of $\mathscr{L}$; but $[0, 1]$ is *not* a member of $\mathscr{L}$. $\mathscr{L}$ consists *only* of those sets specified, namely those which are of the form $(-\infty, a]$ where $a \in \mathsf{R}$.

A lattice $\mathscr{R}$ of subsets of a set X is called a *ring* of sets provided that whenever A and B are members of $\mathscr{R}$, then $A \backslash B$ is a member of $\mathscr{R}$ (thus $\mathscr{R}$ is closed under the operations of $\cup$, $\cap$, and $\backslash$).

Example: Let X be arbitrary and let $\mathscr{R}$ be the collection of all finite subsets of X.

We have already seen that $\mathscr{R}$ is a lattice. If $A \in \mathscr{R}$ and $B \in \mathscr{R}$, then A and B are finite subsets of R; $A \backslash B$ is a subset of A and thus $A \backslash B$ is finite so $A \backslash B \in \mathscr{R}$.

Example: Let $X = \mathsf{R}$ and let $\mathscr{R}$ be the collection of all bounded subsets of X (A is bounded if there exists $M > 0$ such that $x \in A \Rightarrow |x| \leq M$).

Let A and B be members of $\mathscr{R}$, then A and B are bounded subsets of R. $A \cup B$ is bounded by the larger of the bounds for A and B, thus $A \cup B \in \mathscr{R}$. $A \cap B$ and $A \backslash B$ are subsets of A and thus are bounded by the bound of A, so $A \cap B \in \mathscr{R}$ and $A \backslash B \in \mathscr{R}$.

Even though every ring is a lattice, there are some lattices which are not rings. Note that if $\mathcal{R}$ is a ring of sets, then there is some $A \in \mathcal{R}$; thus $\varnothing = A \backslash A \in \mathcal{R}$ so each ring contains the empty set—this is not necessarily true for a lattice. It can be shown that a ring of sets is a ring in the sense of modern algebra under the operations of $\triangle$ and $\cap$, (where

$$A \triangle B = A \cup B \backslash A \cap B).$$

In the case of numbers, the set Q of rational numbers may be sufficient if we are interested in addition, subtraction, multiplication, and division; however, to consider limits of sequences of numbers, we must consider the entire set of real numbers. Thus, we must consider a more complex system. In the same way, we must work within a more complex class of sets in order to consider "limits" of sequences of sets. First, we require a broader concept of union and intersection.

Definition 2: By an *indexed class* of sets $\mathcal{B}$ with non-empty *index set* $\mathfrak{A}$ we shall mean a collection $\mathcal{B}$ of sets with the property that to each α in $\mathfrak{A}$, there corresponds exactly one of the sets of $\mathcal{B}$. For $\alpha \in \mathfrak{A}$ we denote the corresponding set in $\mathcal{B}$ by B_α, and we denote the indexed class $\mathcal{B}$ by $\mathcal{B} = \{B_\alpha \mid \alpha \in \mathfrak{A}\}$.

For example, if $\mathfrak{A} = \{1, 2\}$ then we have only two sets B_1 and B_2. If $\mathfrak{A} = N$, then we have a sequence of sets $\{B_1, B_2, \ldots, B_n, \ldots\}$ or $\{B_n\}_{n=1}^\infty$.

Definition 3: Let $\{B_\alpha \mid \alpha \in \mathfrak{A}\}$ be an indexed class of sets.

(1)   The *union* of the class is

$$\bigcup_\mathfrak{A} B_\alpha = \bigcup \{B_\alpha \mid \alpha \in \mathfrak{A}\} = \{x \mid x \in B_\alpha \text{ for some } \alpha \in \mathfrak{A}\}.$$

(2)   The *intersection* of the class is

$$\bigcap_\mathfrak{A} B_\alpha = \bigcap \{B_\alpha \mid \alpha \in \mathfrak{A}\} = \{x \mid x \in B_\alpha \text{ for each } \alpha \in \mathfrak{A}\}.$$

For a sequence $\{B_n\}_{n=1}^\infty$ of sets, we sometimes use the notations $\bigcup_1^\infty B_n$ and $\bigcap_1^\infty B_n$.

Example: For each $n \in N$, let $B_n = (2 - 1/n, 3 + 1/n]$.

This class of sets is illustrated in figure 1.3.

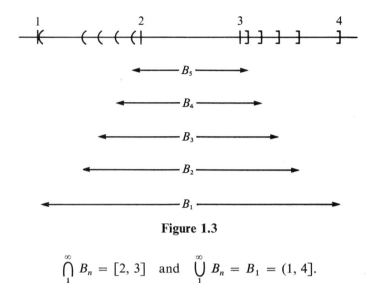

Figure 1.3

$$\bigcap_1^\infty B_n = [2, 3] \quad \text{and} \quad \bigcup_1^\infty B_n = B_1 = (1, 4].$$

Example: For each $n \in N$, let $B_n = (2 - 1/n, 3 - 1/n]$, as illustrated in figure 1.4.

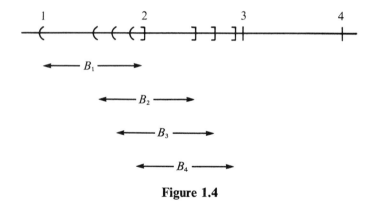

Figure 1.4

Here we have $\bigcap_1^\infty B_n = \{2\}$ and $\bigcup_1^\infty B_n = (1, 3)$.

Example: For each $a \in R$, let $B_a = (-\infty, a]$, then $\bigcap_R B_a = \varnothing$ and $\bigcup_R B_a = R$.

Just as in the preceding section, important relationships exist between these concepts of union and intersection.

EXTENDED DISTRIBUTIVE LAWS: Let A be a set and let $\{B_\alpha \mid \alpha \in \mathfrak{A}\}$ be an indexed class of sets, then

$$(1) \quad A \cap \left(\bigcup_\mathfrak{A} B_\alpha \right) = \bigcup_\mathfrak{A} (A \cap B_\alpha),$$

$$(2) \quad A \cup \left(\bigcap_\mathfrak{A} B_\alpha \right) = \bigcap_\mathfrak{A} (A \cup B_\alpha).$$

Proof: We shall prove (1) and leave (2) for an exercise.

$$x \in A \cap \left(\bigcup_\mathfrak{A} B_\alpha \right) \Leftrightarrow x \in A \quad \text{and} \quad x \in \bigcup_\mathfrak{A} B_\alpha$$

$$\Leftrightarrow x \in A \quad \text{and for some } \alpha \in \mathfrak{A}, \; x \in B_\alpha$$

$$\Leftrightarrow x \in A \cap B_\alpha \quad \text{for some } \alpha \in \mathfrak{A}$$

$$\Leftrightarrow x \in \bigcup_\mathfrak{A} (A \cap B_\alpha). \quad \square$$

EXTENDED DEMORGAN'S LAWS: Let A be a set and let $\{B_\alpha \mid \alpha \in \mathfrak{A}\}$ be an indexed class of sets, then

$$(1) \quad A\backslash \bigcup_\mathfrak{A} B_\alpha = \bigcap_\mathfrak{A} (A\backslash B_\alpha),$$

$$(2) \quad A\backslash \bigcap_\mathfrak{A} B_\alpha = \bigcup_\mathfrak{A} (A\backslash B_\alpha).$$

Proof: Again (2) will be left for an exercise.

$$x \in A\backslash \left(\bigcup_\mathfrak{A} B_\alpha \right) \Leftrightarrow x \in A \quad \text{but} \quad x \notin \bigcup_\mathfrak{A} B_\alpha$$

$$\Leftrightarrow x \in A \quad \text{but for each } \alpha \in \mathfrak{A}, \; x \notin B_\alpha$$

$$\Leftrightarrow x \in A\backslash B_\alpha \quad \text{for each } \alpha \in \mathfrak{A}$$

$$\Leftrightarrow x \in \bigcap_\mathfrak{A} (A\backslash B_\alpha). \quad \square$$

If all sets under consideration are subsets of some X, then a special case of the Extended DeMorgan's Laws is

$$\left(\bigcup_\mathfrak{A} B_\alpha \right)^c = \bigcap_\mathfrak{A} (B_\alpha^c),$$

$$\left(\bigcap_\mathfrak{A} B_\alpha \right)^c = \bigcup_\mathfrak{A} (B_\alpha^c).$$

Definition 4:   A collection $\mathscr{S}$ of subsets of a set X is called a *sigma-ring* (σ-ring) provided

(1) $\mathscr{S}$ is a ring of sets,
(2)   For every sequence $\{A_n\}_1^\infty$ of members of $\mathscr{S}$, we have $\bigcup_1^\infty A_n \in \mathscr{S}$ also.

Example: Let A and B be subsets of a set X, and let

$$\mathscr{S} = \{\varnothing, A, B, A \cap B, A \cup B, A\backslash B, B\backslash A, A \triangle B\}.$$

Then $\mathscr{S}$ is a σ-ring.

It must be checked case by case that $\mathscr{S}$ is a ring of subsets of X and this is not difficult. If $\{A_n\}_1^\infty$ is any sequence of members of $\mathscr{S}$, then there can be only a finite number of different A_n's ($\mathscr{S}$ is finite). Thus $\bigcup_1^\infty A_n$ reduces to a finite union and a simple induction proof shows that any ring is closed under finite unions.

Example: Let X be an arbitrary set and define the *power set* $\mathscr{P}(X)$ to be the collection of all subsets of X. Then $\mathscr{P}(X)$ is a σ-ring.

If A and B are members of $\mathscr{P}(X)$, then each is a subset of X; thus, $A \cup B$, $A \cap B$, and $A\backslash B$ are subsets of X and so they are members of $\mathscr{P}(X)$. If $\{A_n\}_1^\infty$ is a sequence in $\mathscr{P}(X)$, then each A_n is a subset of X and hence $\bigcup_1^\infty A_n \subseteq X$ so $\bigcup_1^\infty A_n \in \mathscr{P}(X)$.

Example: Let $X = \mathsf{R}$ and $\mathscr{S}$ be the collection of all countable subsets of R (recall that a set is countable if it can be put into a one-to-one correspondence with some subset of N). Then $\mathscr{S}$ is a σ-ring.

Let A and B be members of $\mathscr{S}$, then each is a countable subset of R. Since $A \cap B \subseteq A$ and $A\backslash B \subseteq A$, then $A \cap B$ and $A\backslash B$ are countable, therefore $A \cap B \in \mathscr{S}$ and $A\backslash B \in \mathscr{S}$. The set Z of all integers can be put into a one-to-one correspondence with N (namely $n \mapsto 2n$ if $n \geq 0$ and $n \mapsto 2(-n) - 1$ if $n < 0$). From this fact, it follows easily that if A and B are countable, then so is $A \cup B$, therefore $\mathscr{S}$ is closed under $\cup$. So far, we have shown that $\mathscr{S}$ is a ring of subsets of R. Let $\{A_n\}_1^\infty$ be a sequence of members of $\mathscr{S}$; that is, each A_n is countable. The members of A_n can be listed

$$A_n = \{a_{n,1}, a_{n,2}, \ldots, a_{n,m}, \ldots\},$$

where we continue listing some *one* of the elements if A_n is finite. Let $A = \bigcup_1^\infty A_n$; then the elements of A can be listed as in figure 1.5, and we may "count" the elements as shown. Here the correspondence is $1 \leftrightarrow a_{1,1}$, $2 \leftrightarrow a_{1,2}$, $3 \leftrightarrow a_{2,1}$, $4 \leftrightarrow a_{3,1}$, etc. The one problem in this method of counting is that some members of A may be counted more than once. Thus, we do not have A in a one-to-one correspondence with the entire set $\mathbf{N}$, but only some subset of $\mathbf{N}$. However, this is all that the definition of countable requires, hence A is countable and so $\bigcup_1^\infty A_n = A \in \mathscr{S}$ and $\mathscr{S}$ is a σ-ring.

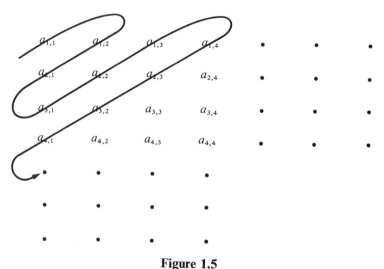

Figure 1.5

In each of the types of collections listed above (lattice, ring, σ-ring), we did not require that X itself be a member of the collection even though the collection was composed of (some of the) subsets of X. If $\mathscr{R}$ is a ring of sets and X itself is a member of $\mathscr{R}$, then $\mathscr{R}$ is called an *algebra* of sets. A σ-ring of subsets of X which contains X itself is called a *sigma algebra* (σ-algebra).

Example: Let X be an arbitrary set. Then $\mathscr{P}(X)$ is a σ-algebra of subsets of X.

Example: Let $X = \mathbf{Q}$ and let $\mathscr{R}$ be the collection of all subsets A of $\mathbf{Q}$ with the property that either A is finite or $X \backslash A$ is finite (in the latter case A is said to be *co-finite*). Then $\mathscr{R}$ is an algebra but not a σ-algebra.

Let A and B be members of $\mathscr{R}$.

Case I: A and B are finite.
            If A and B are finite, then so are $A \cup B$, $A \cap B$, and $A\backslash B$
            as in a previous example, so $A \cup B \in \mathscr{R}$, $A \cap B \in \mathscr{R}$, and
            $A\backslash B \in \mathscr{R}$.

Case II: A and B are co-finite.
            $Q\backslash(A \cup B) = (Q\backslash A) \cap (Q\backslash B)$ which is finite, so $A \cup B$ is
            co-finite   and   hence $A \cup B \in \mathscr{R}$. $Q\backslash(A \cap B) =$
            $(Q\backslash A) \cup (Q\backslash B)$ which is also finite, so $A \cap B \in \mathscr{R}$ since
            it is co-finite.
            $A\backslash B = A \cap (Q\backslash B)$ which is finite, so $A\backslash B \in \mathscr{R}$.

Case III: One (say A) is finite and one, co-finite. $A \cap B \subseteq A$ so
            $A \cap B$ is finite and hence a member of $\mathscr{R}$.
            $A\backslash B \subseteq A$ so $A\backslash B$ is finite. $Q\backslash(B\backslash A) = (Q\backslash B) \cup A$ which
            is finite so $B\backslash A$ is co-finite.
            $Q\backslash(A \cup B) = (Q\backslash A) \cap (Q\backslash B) \subseteq Q\backslash B$ which is finite so
            $A \cup B$ is co-finite.

Thus we have shown that $\mathscr{R}$ is a ring of subsets of Q. Now Q itself is
co-finite since $Q\backslash Q = \varnothing$ which is finite and therefore $Q \in \mathscr{R}$, so $\mathscr{R}$ is
an algebra. $\mathscr{R}$ is not a σ-algebra, however, since for each $n \in N$, $\{n\}$
is a finite set so $\{n\} \in \mathscr{R}$ but $N = \bigcup_{n=1}^{\infty} \{n\}$ is not a member of $\mathscr{R}$ since
it is neither finite nor co-finite.

We shall find that a σ-algebra is sufficiently complex to allow us to
perform all of the "set arithmetic" in which we are interested. It is *not*
essential to memorize all of the other types of collections defined in this
section (lattice, ring, σ-ring, algebra). It *is* essential to realize that each is a
collection of sets in which certain types of set arithmetic are allowed. The
exercises in this section are designed to help gain this realization and to
give experience working with collections of sets.

Exercises:

1. Prove (2) in the Extended Distributive Laws.

2. Prove (2) in the Extended DeMorgan's Laws.

3. In each of (a)–(g), $\mathscr{M}$ is a collection of sets; determine which of the
    following statements is (are) true: (i) $\mathscr{M}$ is closed under $\cup$, (ii) $\mathscr{M}$ is

closed under $\cap$, (iii) $\mathcal{M}$ is closed under $\backslash$, (iv) if $\{A_n\}_1^\infty$ is a sequence in $\mathcal{M}$, then $\bigcap_1^\infty A_n \in \mathcal{M}$, (v) if $\{A_n\}_1^\infty$ is a sequence in $\mathcal{M}$, then $\bigcup_1^\infty A_n \in \mathcal{M}$, (vi) $\mathcal{M}$ is a σ-algebra.

(a) $\mathcal{M} = \{(a, \infty) \mid a \in \mathsf{R}\}$.

(b) $\mathcal{M} = \{(a, b] \mid a, b \in \mathsf{R}, a \le b\}$.

(c) $\mathcal{M} = \{A \mid A \subseteq \mathsf{R} \text{ and either } A \text{ is countable or } \mathsf{R}\backslash A \text{ is countable}\}$.

(d) $\mathcal{M} = \{\varnothing, \mathsf{R}\}$.

(e) $\mathcal{M} = \{A, B, A \cup B, A \cap B\}$ where $A \subseteq \mathsf{R}$ and $B \subseteq \mathsf{R}$ are fixed.

(f) $\mathcal{M}$ is the collection of all bounded subsets of R.

(g) $\mathcal{M} = \{[0, a] \mid a \in \mathsf{R}$ and $0 \le a \le 1\}$.

4. Suppose $\mathcal{R}$ is a non-empty collection of subsets of X such that
(a) $A \in \mathcal{R}, B \in \mathcal{R} \Rightarrow A \cup B \in \mathcal{R}$
(b) $A \in \mathcal{R}, B \in \mathcal{R} \Rightarrow A \backslash B \in \mathcal{R}$.
Prove that $\mathcal{R}$ is a ring of sets.

5. Suppose that $\mathcal{R}$ is a non-empty collection of subsets of X such that
(a) $A \in \mathcal{R}, B \in \mathcal{R} \Rightarrow A \cap B \in \mathcal{R}$
(b) $A \in \mathcal{R} \Rightarrow A^c = X \backslash A \in \mathcal{R}$.
Prove that $\mathcal{R}$ is an algebra of subsets of X.

6. Prove that every finite ring is a σ-ring.

7. $\mathcal{R}$ is a non-empty collection of subsets of X such that
(a) If $\{A_n\}_1^\infty$ is any sequence in $\mathcal{R}$, then $\bigcup_1^\infty A_n \in \mathcal{R}$
(b) $A \in \mathcal{R}, B \in \mathcal{R} \Rightarrow A \backslash B \in \mathcal{R}$.
Prove that $\mathcal{R}$ is a σ-ring.

*8. $\mathcal{R}$ is a non-empty collection of subsets of X such that
(a) If $\{A_n\}_1^\infty$ is any sequence in $\mathcal{R}$, then $\bigcap_1^\infty A_n \in \mathcal{R}$
(b) $A \in \mathcal{R} \Rightarrow A^c = X \backslash A \in \mathcal{R}$.
Prove that $\mathcal{R}$ is a σ-algebra of subsets of X.

9. Let $\mathcal{M} = \{(a, b] \mid a, b \in \mathsf{R}, a \le b\}$ and let $\mathcal{R}$ be the collection of all subsets of R which can be written as a finite union of members of $\mathcal{M}$. For example, each of the following is a member of $\mathcal{R}$:
$(0, 3], (1, \pi] \cup (4, 6], (-3/2, 1] \cup (2, 4] \cup (3, 7]$.
Prove each of the following:
(a) Each member of $\mathcal{R}$ can be written as a union of a finite *disjoint* collection of members of $\mathcal{M}$.
(b) $\mathcal{R}$ is closed under $\cup$

(c) $\mathcal{M}$ is closed under $\cap$

(d) $\mathcal{R}$ is closed under $\cap$

(e) If $A \in \mathcal{M}$ and $B \in \mathcal{M}$, then $A \backslash B \in \mathcal{R}$

(f) $\mathcal{R}$ is closed under $\backslash$

(g) $\mathcal{R}$ is a ring of subsets of R.

10. $\mathcal{R}$ is an algebra with only a finite number of members, say n. Prove that n is even.

11. Prove that Q is countable. (Hint: see figure 1.5.)

12. Find $\bigcap_1^\infty B_n$ and $\bigcup_1^\infty B_n$ in each of the following

(a) $B_n = (a - 1/n, b + 1/n)$ for $n \in$ N, $(a < b)$

(b) $B_n = (a - 1/n, a)$ for $n \in$ N

(c) $B_n = (a - 1/n, a]$ for $n \in$ N

(d) $B_n = (n, n + 1]$ for $n \in$ N

(e) $B_n = (1/n + 1, 1/n]$ for $n \in$ N.

*13. Show that each σ-algebra, Σ, is closed under countable intersections, i.e., if $\{A_n\}_1^\infty$ is a sequence in Σ, then $\bigcap_1^\infty A_n \in \Sigma$ also.

2

Probability Measure Spaces

2.1 Basic Definitions

As was mentioned in chapter 1, probability is concerned with the assignment of numbers to occurrences and the mathematical model for an occurrence is a set. Any function that assigns numbers to sets is called a *set function*. For example, for a finite set A, we shall let $\#(A)$ be the number of elements in A (we shall say that A is *finite* if it is the empty set or if it can be put in one-to-one correspondence with $\{1, 2, \ldots, n\}$ for some $n \in N$). For instance, $\#(\varnothing) = 0$, $\#(\{1, 7, 6\}) = 3$, etc.). Let $\mathscr{R}$ be the ring of all finite subsets of a set X; then $\#$ is a function from $\mathscr{R}$ into the set $[0, \infty)$ of non-negative real numbers. The set function $\#$ has certain properties that are important enough to point out.

(1) If $A \in \mathscr{R}$ and $B \in \mathscr{R}$ and $A \cap B = \varnothing$, then $\#(A \cup B) = \#(A) + \#(B)$ ($\#$ is said to be 2-additive).

(2) If $\{A_1, \ldots, A_n\} \subseteq \mathscr{R}$ for any $n \in \mathsf{N}$ and $A_i \cap A_j = \varnothing$ whenever $i \neq j$, then $\#(A_1 \cup \cdots \cup A_n) = \#(A_1) + \cdots + \#(A_n)$ ($\#$ is said to be *finitely additive*).

(3) If $A \in \mathscr{R}$ and $B \in \mathscr{R}$ and $A \subseteq B$, then $\#(B \backslash A) = \#(B) - \#(A)$ ($\#$ is said to be *subtractive*).

(4) If $A \in \mathscr{R}$ and $B \in \mathscr{R}$ and $A \subseteq B$, then $\#(A) \leq \#(B)$ ($\#$ is said to be *monotone* or *non-decreasing*).

(5) If $A \in \mathscr{R}$ and $B \in \mathscr{R}$, then $\#(A \cup B) = \#(A) + \#(B) - \#(A \cap B)$ ($\#$ is said to be *modular*).

Property (1) should be clear and property (2) follows by a simple induction argument. Property (3) follows from (1) since $B = (B \backslash A) \cup A$ and $(B \backslash A) \cap A = \varnothing$. Property (4) follows from (3) and the fact that $\#(B \backslash A) \geq 0$. To prove (5), we note that $A \cup B = A \cup (B \backslash A) = A \cup (B \backslash (A \cap B))$ and $A \cap (B \backslash (A \cap B)) = \varnothing$. Thus, $\#(A \cup B) = \#(A) + \#(B \backslash (A \cap B)) = \#(A) + \#(B) - \#(A \cap B)$.

We shall use $\#$ later, but for the moment we will note simply that properties (2)–(5) follow from (1) and the fact that $\#(A) \geq 0$ for each $A \in \mathscr{R}$. Thus any non-negative set function which satisfies property (1) will also satisfy properties (2)–(5). In particular, we shall see that this is true for a probability measure.

Definition 1:   Let S be a non-empty set and Σ be a σ-algebra of subsets of S. Let P be a set function from Σ into the set of real numbers with the properties:

(a) for each $E \in \Sigma$, $P(E) \geq 0$,

(b) P is 2-additive: if $A \in \Sigma$ and $B \in \Sigma$ and $A \cap B = \varnothing$ then $P(A \cup B) = P(A) + P(B)$,

(c) P is *countably additive*: if $\{A_n\}_1^\infty$ us a sequence in Σ such that $A_i \cap A_j = \varnothing$ whenever $i \neq j$, then

$$P\left(\bigcup_1^\infty A_n\right) = \sum_1^\infty P(A_n),$$

(d) $P(S) = 1$.

Then (S, Σ, P) is called a *probability measure space* (abbreviated p.m.s.); S is called the *sample space*; members of Σ are called *events*; and P is called a *probability measure*. $P(E)$ is usually read "the probability of E."

The following sections will exhibit many examples of probability measure spaces; in the remainder of this section we shall explore some properties which all probability measure spaces share.

THEOREM 1:   Let (S, Σ, P) be a probability measure space.

(1) $P(\emptyset) = 0$.

(2)   If $\{E_i\}_1^n$ is a pairwise disjoint collection in Σ ($E_i \cap E_j = \emptyset$ when $i \neq j$) then

$$P\left(\bigcup_1^n E_i\right) = \sum_1^n P(E_i).$$

(3)   If $E \in \Sigma$ and $F \in \Sigma$ and $E \subseteq F$, then $P(F\backslash E) = P(F) - P(E)$.

(4)   If $E \in \Sigma$ and $F \in \Sigma$ and $E \subseteq F$, then $P(E) \leq P(F)$.

(5)   If $E \in \Sigma$ and $F \in \Sigma$, then

$$P(E \cup F) = P(E) + P(F) - P(E \cap F).$$

(6)   If $E \in \Sigma$, then $P(E^c) = P(S\backslash E) = 1 - P(E)$.

Proof: (1) $\emptyset = \emptyset \cup \emptyset$ and $\emptyset \cap \emptyset = \emptyset$, thus by (b) in the definition of a probability measure space, $P(\emptyset) = P(\emptyset \cup \emptyset) = P(\emptyset) + P(\emptyset)$ so $P(\emptyset) = 0$.

Properties (2)–(5) follow just as did the corresponding properties for $\#$. (6): From property (3), $P(S\backslash E) = P(S) - P(E) = 1 - P(E)$. $\square$

So far, property (c) in the definition of a p.m.s. has not been mentioned. In the case of $\#$, this situation arises only in a trivial manner: if $\bigcup_1^{\infty} A_n$ is finite and $A_i \cap A_j = \emptyset$ whenever $i \neq j$, then it must be the case that $A_i = \emptyset$ for all but at most a finite number of A_i's. Hence the countable union may be considered as a finite union and finite additivity gives the required result. Notice, however, that the collection of finite subsets of a set is only a ring and not a σ-algebra; in a σ-algebra Σ, we may very well have some non-trivial, countable, pairwise-disjoint collection in Σ and the union *must* be in Σ. Of course if Σ contains only a finite number of different sets, then an argument similar to that with $\#$ will automatically give countable additivity from 2-additivity.

Countable additivity may be considered an extension of finite additivity, but perhaps equally strong reasons for specifying this property are the concepts of continuity which follow from it.

THEOREM 2:   Let (S, Σ, P) be a probability measure space.

(a) Let $A \in \Sigma$ and $\{E_n\}_1^\infty$ be a sequence in Σ such that $E_1 \subseteq E_2 \subseteq \cdots \subseteq E_n \subseteq E_{n+1} \subseteq \cdots$ and $A = \bigcup_1^\infty E_n$. Then $P(A) = \lim_{n \to \infty} P(E_n)$ (P is *continuous from below* at A).

(b) Let $A \in \Sigma$ and $\{E_n\}_1^\infty$ be a sequence in Σ such that $E_1 \supseteq E_2 \supseteq \cdots \supseteq E_n \supseteq E_{n+1} \supseteq \cdots$ and $A = \bigcap_1^\infty E_n$. Then $P(A) = \lim_{n \to \infty} P(E_n)$ (P is *continuous from above* at A).

(Note that in each case, it is reasonable to write

$$P(\lim_{n \to \infty} E_n) = \lim_{n \to \infty} P(E_n),$$

hence the term continuity.)

Proof: (a) Let $B_1 = E_1$ and for $n > 1$, let $B_n = E_n \backslash E_{n-1}$. If $i \neq j$, then $B_i \cap B_j = \emptyset$; also $\bigcup_1^\infty B_n = \bigcup_1^\infty E_n = A$. By countable additivity,

$$P(A) = \sum_1^\infty P(B_n) = \sum_{n=2}^\infty P(B_n) + P(B_1)$$

$$= \sum_{n=2}^\infty P(E_n \backslash E_{n-1}) + P(E_1)$$

$$= \lim_{N \to \infty} \sum_{n=2}^N (P(E_n) - P(E_{n-1})) + P(E_1)$$

$$= \lim_{N \to \infty} P(E_N) - P(E_1) + P(E_1)$$

$$= \lim_{N \to \infty} P(E_N).$$

(b)   For each $n \in \mathbb{N}$, let $B_n = E_1 \backslash E_n$; then $\varnothing = B_1 \subseteq B_2 \subseteq \cdots$ $\subseteq B_n \subseteq B_{n+1} \subseteq \cdots$ and $\bigcup_1^\infty B_n = \bigcup_1^\infty (E_1 \backslash E_n) = E_1 \backslash \bigcup_1^\infty E_n = E_1 \backslash A$. Thus by (a),

$$P(E_1) - P(A) = P(E_1 \backslash A) = \lim_{n \to \infty} P(B_n)$$

$$= \lim_{n \to \infty} P(E_1 \backslash E_n)$$

$$= \lim_{n \to \infty} [P(E_1) - P(E_n)]$$

$$= P(E_1) - \lim_{n \to \infty} P(E_n).$$

Then (b) follows immediately.   □

Before ending this section on basic concepts and going on to specific examples of probability measure spaces, it might be worthwhile to investigate just how a probability measure space should be set up to reflect some real-life process or experiment. Suppose our real-life experiment is throwing a pair of dice. There are several types of questions that we may ask about the outcome. Each of these may dictate a different probability model—and we emphasize that a probability measure space *is* a model and, as such, can be neither "right" nor "wrong," but either a good predictor or a bad one (and, of course, there are various degrees of quality of prediction).

Case I: Suppose we are interested in what is on each die (singular of dice). Then we could take, as a sample space, all pairs: $S = \{1\text{-}1, 1\text{-}2, \ldots, 6\text{-}5, 6\text{-}6\}$. We must then pick a collection of events Σ and a reasonable probability measure P.

Case II: Suppose we wish to know only the sum of the two numbers on the dice. Then a reasonable sample space may be $S = \{2, 3, 4, \ldots, 12\}$.

Case III: Suppose we wish only to know whether the sum is even or odd. A sufficient sample space may be $S = \{E$ (for even), O (for odd)$\}$.

Of course, each of these choices of a sample space requires a corresponding choice of events and probability measure. The choice of the sample space depends largely on the types of questions that are to be answered.

Exercises:

1. Derive the fact that $P(\varnothing) = 0$ using only property (c) of the properties in the definition of a p.m.s.

2. Derive property (b) from property (c) in the definition of a p.m.s.

3. Let Σ be a σ-algebra of subsets of a non-empty set S and P be a function from Σ into R such that properties (a), (b), and (d) in the definition of a p.m.s. are satisfied. Prove that these are equivalent:
   (1) P is countably additive.
   (2) P is continuous from below at each member of Σ.
   (3) P is continuous from above at each member of Σ.
   (4) P is continuous from above at $\varnothing$.
   (5) P is continuous from below at S.

4. Let Σ be a σ-algebra of subsets of a non-empty set S and let P_1 and P_2 be probability measures on Σ. Let $0 \leq \lambda \leq 1$, and $P(E) = \lambda P_1(E) + (1 - \lambda)P_2(E)$ for $E \in \Sigma$ (P is said to be a *convex combination* of P_1 and P_2). Prove that P is a probability measure on Σ.

*5. Let (S, Σ, P) be a p.m.s. and $C \in \Sigma$ such that $P(C) > 0$. Then the function P_C defined by

$$P_C(E) = \frac{P(E \cap C)}{P(C)}$$

   is a probability measure on Σ.

6. (Finite subadditivity) Let (S, Σ, P) be a p.m.s. and $\{E_i\}_1^n$ any collection in Σ. Then

$$P\left(\bigcup_1^n E_i\right) \leq \sum_1^n P(E_i).$$

7. (Countable subadditivity) Let (S, Σ, P) be a p.m.s. and $\{E_i\}_1^\infty$ any sequence in Σ. Then

$$P\left(\bigcup_1^\infty E_i\right) \leq \sum_1^\infty P(E_i).$$

8. (Finite modularity) Let (S, Σ, P) be a p.m.s. and $\{E_i\}_1^n$ any collection in Σ.

(a)   Prove that

$$P(E_1 \cup E_2 \cup E_3)$$
$$= P(E_1) + P(E_2) + P(E_3)$$
$$- (P(E_1 \cap E_2) + P(E_1 \cap E_3) + P(E_2 \cap E_3))$$
$$+ P(E_1 \cap E_2 \cap E_3)$$

(b)   Derive and prove a similar formula for $P(E_1 \cup E_2 \cup \cdots \cup E_n)$.

9.   Let (S, Σ, P) be a p.m.s. and $\{E_i\}_1^\infty$ a sequence in Σ with the property that if $i \neq j$ then $P(E_i \cap E_j) = 0$. (Caution: $P(E_i \cap E_j) = 0 \not\Rightarrow E_i \cap E_j = \varnothing$). Prove

$$P\left(\bigcup_1^\infty E_i\right) = \sum_1^\infty P(E_i).$$

10.   Suppose (S, Σ, P) is a p.m.s. and $S = \{a, b, c\}$. Suppose that $\Sigma = \mathscr{P}(S)$ and that $P(\{a, b\}) = 1/2 = P(\{b, c\})$. Find $P(\{a\})$, $P(\{b\})$, and $P(\{c\})$.

2.2   Discrete Probability Measure Spaces

Very often in considering a random process or experiment, the various outcomes may seem to have equal likelihood. For example, in throwing a fair die, one is just as likely to obtain a 1 as a 5. We now exhibit a probability measure space which models this situation.

Definition 1:   Let S be a non-empty finite set and $\Sigma = \mathscr{P}(S)$ (the collection of all subsets of S). Define P on Σ by

$$P(A) = \frac{\#(A)}{\#(S)} = \frac{\text{number of elements in } A}{\text{number of elements in } S}.$$

Then P is called the *equiprobable measure on* Σ.

By considering the corresponding properties of $\#$, it is clear that properties (a), (b), and (d) in the definition of a p.m.s. are satisfied. Property (c) (countable additivity) is satisfied trivially since every countable pairwise-disjoint union of members of Σ must reduce to a finite union.

Another way of approaching the equiprobable measure is to begin with the set S of n elements and assign each element s, a weight of $w(s) = 1/n$. Then for each event, we define

$$P(A) = \sum_{s \in A} w(s).$$

Example: A fair die is tossed.

Let $S = \{1, 2, 3, 4, 5, 6\}$ and assign weights $w(s) = 1/6$ for each $s \in S$. Let E be the event $E = \{2, 4, 6\}$; we may write $P(\text{even})$ for $P(E)$ in this case; then $P(\text{even}) = 1/6 + 1/6 + 1/6 = 1/2$.

If a process or experiment has n possible outcomes and the statement is made that one occurs "at random," then in the absence of additional information we shall assume that each element is equally likely and hence is assigned a weight of $1/n$.

In the preceding example we could have taken $S = \text{R}$ and assigned weights as follows:

$$w(s) = \begin{cases} 1/6 & \text{if } s = 1, 2, 3, 4, 5, \text{ or } 6 \\ 0 & \text{if not} \end{cases}$$

Then for any σ-algebra Σ of subsets of R, we can define

$$P(A) = \sum_{s \in A} w(s) \quad \text{for} \quad A \in \Sigma \left(\sum_{s \in \emptyset} w(s) = 0 \right).$$

The sum makes sense since $w(s) = 0$ for all but a finite number of s's. This observation motivates the following definition.

Definition 2: Let S be a non-empty set and Σ a σ-algebra of subsets of S. Let w be a function from S into the set of real numbers such that

(a) $w(s) \geq 0$ for each $s \in S$,

(b) there is a countable subset C of S such that if $s \notin C$ then $w(s) = 0$,

(c) $\sum_{s \in S} w(s) = 1$.

For $A \in \Sigma$, define $P(A) = \sum_{s \in A} w(s)$ $(P(\emptyset) = 0)$. Then (S, Σ, P) is called a *discrete probability measure space* and w is called a *weight function*.

To prove that a discrete probability measure space is actually a probability measure space, we must show that P, as defined above, is a probability measure; we shall need the following lemma.

Lemma: Let $\{a_i\}_1^\infty$ be a sequence of real numbers such that

$$\sum_1^\infty |a_i| < \infty \quad \left(\sum_1^\infty a_i \text{ converges absolutely}\right).$$

Then if $i \leftrightarrow j_i$ is any rearrangement of the indices (a rearrangement is a one-to-one correspondence between N and N) then $\sum_{i=1}^\infty a_{j_i}$ converges to $\sum_{i=1}^\infty a_i$ and $\sum_{i=1}^\infty |a_{j_i}| < \infty$.

Proof: Let $\varepsilon > 0$, then since $\sum_{k=1}^\infty |a_k| < \infty$, there exists a positive integer N such that $\sum_{k=N}^\infty |a_k| < \varepsilon$. Now there exists only a finite set of integers less than N so let $M = \max \{i \mid j_i < N\}$; then for $j_i < N$ we have $i \le M$. If $n > M$, then

$$\left|\sum_{k=1}^\infty a_k - \sum_{i=1}^n a_{j_i}\right| = \left|\sum_{\substack{k \ne j_i \\ \text{for } i=1,2,\ldots,n}} a_k\right| \le \sum_{k \ge N} |a_k| < \varepsilon.$$

Thus the sequence

$$\left\{\sum_{i=1}^n a_{j_i}\right\}$$

converges to $\sum_{k=1}^\infty a_k$. To show that $\sum_{i=1}^n a_{j_i}$ converges absolutely we consider the special case in which all the terms of the sequence are non-negative. □

If w is a weight function, then $\sum_{s \in A} w(s)$ is now well defined since there is only a countable collection of s's in A such that $w(s) > 0$ and since the *series* $\sum_{s \in C} w(s)$ converges absolutely.

Proposition 1:   *The set function P defined in Definition 2 is a probability measure.*

Proof: Let the countable set C in Definition 2 be

$$C = \{s_1, s_2, \ldots, s_n, \ldots\}.$$

(If C is finite, the properties of a probability measure follow easily.)

(a) We know $w(s) \geq 0$ for each $s \in S$, so if $A \neq \varnothing$, $P(A) = \sum_{s \in A} w(s) \geq 0$; if $A = \varnothing$, $P(A) = 0$. Thus $P(A) \geq 0$ for each $A \in \Sigma$.

(b) Let E and F be disjoint events. Then

$$\{w(s) \mid s \in E \cup F\} = \{w(s) \mid s \in E\} \cup \{w(s) \mid s \in F\}.$$

Thus

$$\sum_{s \in E \cup F} w(s) = \sum_{s \in E} w(s) + \sum_{s \in F} w(s)$$

since the latter sum is simply a rearrangement of the terms in the first sum. Thus $P(E \cup F) = P(E) + P(F)$.

(c) Countable additivity is again a consequence of arrangement–invariance. If $\{E_i\}_1^\infty$ is a pairwise–disjoint sequence in Σ and $E = \bigcup_1^\infty E_i$, then

$$P(E) = \sum_{s \in E} w(s) = \sum_{s \in \bigcup_1^\infty E_i} w(s) = \sum_{s \in E_1} w(s) + \sum_{s \in E_2} w(s) + \cdots$$

$$= \sum_{i=1}^\infty \left(\sum_{s \in E_i} w(s) \right) = \sum_{i=1}^\infty P(E_i).$$

(d) $P(S) = \sum_{s \in S} w(s) = 1.$ $\square$

Example: A die is to be tossed; the die is weighted so that the probability of obtaining an integer 1, 2, 3, 4, 5, or 6 is proportional to that integer (e.g. 4 is twice as likely as 2).

A reasonable sample space is $S = \{1, 2, 3, 4, 5, 6\}$ with $\Sigma = \mathscr{P}(S)$. We assign weights as follows:

$$w(1) = a, \; w(2) = 2a, \; w(3) = 3a, \ldots, \; w(6) = 6a.$$

Since $\sum_{s \in S} w(s) = a + 2a + \cdots + 6a = 21a = 1$, then $a = 1/21$; thus $w(1) = 1/21, w(2) = 2/21, \ldots, w(6) = 6/21$. Now we may compute probabilities, for example, $P(\text{even}) = 2/21 + 4/21 + 6/21 = 12/21$.

Example: A pair of fair dice is tossed and we are interested in the sum of the numbers on the faces.

Here we take as a sample space, $S = \{2, 3, \ldots, 12\}$ and $\Sigma = \mathscr{P}(S)$ and assign weights as follows:

$$w(2) = 1 \cdot a \quad (2 = 1 + 1)$$

$$w(3) = 2 \cdot a \quad (3 = 2 + 1 = 1 + 2)$$

$$w(4) = 3 \cdot a \quad (4 = 3 + 1 = 2 + 2 = 1 + 3)$$

$$w(5) = 4 \cdot a \quad (5 = 4 + 1 = 3 + 2 = 2 + 3 = 1 + 4)$$

$$\vdots$$

$$w(12) = 1 \cdot a \quad (12 = 6 + 6)$$

Now $\sum_{s \in S} w(s) = a + 2a + \cdots + 2a + a = 36a = 1$ so $a = 1/36$. Hence, $w(2) = 1/36, w(3) = 2/36$, etc.; for example $P(\text{divisible by 3}) = P(\{3, 6, 9, 12\}) = 12/36 = 1/3$.

Many more examples of discrete probability measure spaces will be treated later after we have considered some counting formulae and independent repeated trials.

Exercises:

1. In the proof of part (b) of Proposition 1, it is stated that

$$\sum_{s \in E \cup F} w(s) = \sum_{s \in E} w(s) + \sum_{s \in F} w(s)$$

by the lemma concerning rearrangement of series. Justify this statement.

2. In a particular high school, there are 110 juniors and each is required to take a foreign language—either French or Spanish. The French teacher reports an enrollment of 60 while the Spanish teacher reports 75 (French and Spanish are not taught at the same hour!).

- (a) How many students are taking both languages?
- (b) How many take French but not Spanish?
- (c) How many take Spanish but not French?
- (d) A student is chosen at random from the junior class; what is the probability that he is enrolled for exactly one language course?

3. A card is drawn at random from a standard deck of fifty-two. Find the probability of drawing
   - (a) a face card
   - (b) a club face card
   - (c) a club
   - (d) either a club or a face card.

4. Two fair dice are thrown. What is the probability that the sum on the faces shown is divisible by 4?

5. Two fair dice are thrown. What is the probability that the difference of the faces shown is divisible by 3?

6. Two fair dice are thrown; one is regular but the other has two 1's, two 2's, and two 3's. What is the probability that the sum of the numbers obtained is divisible by 3?

7. A student is taking math and French and figures that the probability of his passing at least one of these is 0.8 but that the probability of passing both is 0.1. He figures that he is twice as likely to pass French as he is to pass math. What is the probability that he will pass math?

2.3   Other Probability Measure Spaces

In this section we shall deal with probability measures defined on some common subsets of R. Let Σ be a σ-algebra of subsets of R such that for each $a \in$ R, the set $(-\infty, a]$ is a member of Σ. Then most other familiar sets will also be members of Σ. For example, each set of the form $(a, b]$ with $a < b$ is a member of Σ since

$$(-\infty, b], (-\infty, a] \in \Sigma \Rightarrow (a, b] = (-\infty, b]\backslash(-\infty, a] \in \Sigma.$$

Every singleton set $\{a\}$ is a member of Σ since for each $n \in$ N,

$$(a - 1/n, a] \in \Sigma; \{a\} = \bigcap_1^\infty (a - 1/n, a] \in \Sigma$$

because Σ is closed under countable intersection (why?). Other examples of members of Σ are

$$(a, b) = (a, b]\backslash\{b\}$$

$$(-\infty, a) = (-\infty, a]\backslash\{a\}$$

$$(a, \infty) = \mathsf{R}\backslash(-\infty, a]$$

$$[a, b) = (a, b) \cup \{a\}.$$

The set Q of rational numbers is also a member of Σ since it is a countable union of members of Σ; namely, $\mathsf{Q} = \bigcup \{\{r\} \mid r \in \mathsf{Q}\}$ and Q is countable.

Now certainly $\mathscr{P}(\mathsf{R})$ is one σ-algebra which contains all intervals of the form $(-\infty, a]$ but there may be others. There will always exist a smallest such σ-algebra.

THEOREM 1:   Let $\mathscr{M}$ be any non-empty collection of subsets of a set X. Then there exists a σ-algebra $\Sigma_{\mathscr{M}}$ containing all members of $\mathscr{M}$ and with the property that if Σ is *any* σ-algebra containing all members of $\mathscr{M}$, then $\Sigma_{\mathscr{M}} \subseteq \Sigma$.

Proof: Let $\mathscr{N} = \{\Sigma_\alpha \mid \alpha \in \mathfrak{A}\}$ be the collection of all σ-algebras of subsets of X which contain $\mathscr{M}$ (there is at least one such σ-algebra, namely $\mathscr{P}(X)$). Let $\Sigma_{\mathscr{M}} = \{A \subseteq X \mid A \in \Sigma_\alpha \text{ for each } \alpha \in \mathfrak{A}\}$; we shall first show that $\Sigma_{\mathscr{M}}$ is a σ-algebra containing $\mathscr{M}$. If $\{A_n\}_1^\infty$ is a sequence in $\Sigma_{\mathscr{M}}$, then for each $\alpha \in \mathfrak{A}$, $\{A_n\}_1^\infty$ is a sequence in Σ_α. Now each Σ_α is a σ-algebra so $\bigcup_1^\infty A_n \in \Sigma_\alpha$; thus $\bigcup_1^\infty A_n \in \Sigma_{\mathscr{M}}$. If $A \in \Sigma_{\mathscr{M}}$ then $A \in \Sigma_\alpha$ for each $\alpha \in \mathfrak{A}$; therefore $A^c = X\backslash A \in \Sigma_\alpha$ for each $\alpha \in \mathfrak{A}$. Hence $A^c \in \Sigma_{\mathscr{M}}$ so $\Sigma_{\mathscr{M}}$ is a σ-algebra. Now if $A \in \mathscr{M}$, then $A \in \Sigma_\alpha$ for each $\alpha \in \mathfrak{A}$, so $A \in \Sigma_{\mathscr{M}}$ also. Suppose Σ is some σ-algebra of subsets of X such that $\mathscr{M} \subseteq \Sigma$. Then $\Sigma \in \mathscr{N}$ so $\Sigma = \Sigma_\beta$ for some $\beta \in \mathfrak{A}$. Now whenever $A \in \Sigma_{\mathscr{M}}$ we have that $A \in \Sigma_\alpha$ for each $\alpha \in \mathfrak{A}$, so in particular, $A \in \Sigma_\beta = \Sigma$; thus $\Sigma_{\mathscr{M}} \subseteq \Sigma$. $\square$

We shall call the smallest σ-algebra containing all sets of the form $(-\infty, a]$, the *Borel σ-algebra* of R and denote it by $\mathscr{B}(\mathsf{R})$. The members of $\mathscr{B}(\mathsf{R})$ are called *Borel sets* in R. There *do* exist subsets of R which are not Borel sets, but the construction and exhibition of such a set is difficult.

There are many equivalent ways in which the Borel σ-algebra could have been defined.

THEOREM 2: $\mathscr{B}(\mathsf{R})$ is

(a)   the smallest σ-algebra containing all sets of the form $(a, b]$,
(b)   the smallest σ-algebra containing all sets of the form (a, b),

(c) the smallest σ-algebra containing all sets of the form $[a, b)$,
(d) the smallest σ-algebra containing all sets of the form $[a, b]$,
(e) the smallest σ-algebra containing all sets of the form $(-\infty, a)$,
(f) the smallest σ-algebra containing all sets of the form $[a, \infty)$,
(g) the smallest σ-algebra containing all sets of the form (a, ∞).

The proof is simply an exercise in the use of the properties which a σ-algebra possesses. Each type of set listed above can be "built" from each other type by using σ-algebra properties.

Suppose $(R, \mathscr{B}(R), P)$ is a probability measure space; that is, P is a probability measure defined on the Borel sets of R. Then in particular $P((-\infty, a])$ is defined for each real number a. Consider the function F from R into R defined by

$$F(a) = P((-\infty, a]).$$

F is called the (cumulative) *distribution function* for the probability measure P. All distribution functions share certain properties.

Proposition 1: *Let F be the distribution function for the probability measure P on $\mathscr{B}(R)$.*

(a) If $a \leq b$, then $F(a) \leq F(b)$
(b) $\lim\limits_{a \to -\infty} F(a) = 0$
(c) $\lim\limits_{a \to \infty} F(a) = 1$
(d) F is continuous from the right on R (i.e., $\lim\limits_{x \to a^+} F(x) = F(a)$ for each $a \in$ R.)

Proof: (a) If $a \leq b$, then $(-\infty, a] \subseteq (-\infty, b]$ so $F(a) = P((-\infty, a]) \leq P((-\infty, b]) = F(b)$.
(b) Let $E_n = (-\infty, -n]$ for each $n \in$ N, then $E_1 \supseteq E_2 \supseteq \cdots$ and $\bigcap_1^\infty E_n = \varnothing$. By continuity from above at $\varnothing$,

$$0 = P(\varnothing) = \lim\limits_{n \to \infty} P(E_n) = \lim\limits_{n \to \infty} F(-n),$$

then (b) follows from (a).
(c) Let $E_n = (-\infty, n]$ for $n \in$ N, then $E_1 \subseteq E_2 \subseteq \cdots$ and $\bigcup_1^\infty E_n = $ R. By continuity from below at R,

$$1 = P(R) = \lim\limits_{n \to \infty} P(E_n) = \lim\limits_{n \to \infty} F(n),$$

and again the desired result follows from (a).

(d) Let $a \in R$, then $(-\infty, a] = \bigcap_{1}^{\infty} (-\infty, a + 1/n]$ and if $E_n = (-\infty, a + 1/n]$ for $n \in N$, then $E_1 \supseteq E_2 \supseteq \cdots$. Thus

$$F(a) = P((-\infty, a]) = \lim_{n \to \infty} P(E_n) = \lim_{n \to \infty} F(a + 1/n).$$

We invoke (a) again to ensure that F be continuous from the right at a. $\square$

What is perhaps more interesting than the fact that all distribution functions have these properties is the fact that any function which has these four properties *is* the distribution function for some probability measure on $\mathscr{B}(R)$. (If it looks like a duck, and acts like a duck, and sounds like a duck, then . . . .) Unfortunately, we do not have the machinery necessary to prove this result (see, for example, *A Course in Probability Theory* by K. L. Chung, published by Harcourt, Brace, and World, 1968) but we shall state it.

THEOREM 3:   Let F be a function from R into R. These are equivalent:

(a) F is the distribution function for some probability measure on $\mathscr{B}(R)$.
(b) F satisfies the properties (a), (b), (c), (d) of Proposition 1.

The statement of the theorem above gives no indication of the construction of the probability measure that a distribution function is associated with; however, we can compute probabilities of most of the sets in which we are interested simply by knowing that $P((-\infty, a]) = F(a)$ for each $a \in R$.

THEOREM 4:   If F is the distribution function for a probability measure P on $\mathscr{B}(R)$, then

(a) $P((a, b]) = F(b) - F(a)$ if $a \leq b$

(b) $P((-\infty, b)) = F(b^-) = \lim_{x \to b^-} F(x)$

(c) $P((a, b)) = F(b^-) - F(a)$ if $a \leq c$

(d) $P([a, b]) = F(b) - F(a^-)$ if $a \leq b$

(e) $P([a, b)) = F(b^-) - F(a^-)$ if $a \leq b$

(f) $P((a, \infty)) = 1 - F(a)$

(g) $P([a, \infty)) = 1 - F(a^-)$

(h) $P(\{a\}) = F(a) - F(a^-)$.

Proof: Most will be left as exercises.

(a) $P((a, b]) = P((-\infty, b]\backslash(-\infty, a])$
$$= P((-\infty, b]) - P((-\infty, a]) = F(b) - F(a).$$

(b)   For each $n \in \mathbf{N}$, let $A_n = (-\infty, b - 1/n]$, then $A_1 \subseteq A_2 \subseteq \cdots$

and $\bigcup_1^\infty A_n = (-\infty, b)$.   Thus $P((-\infty, b)) = \lim_{n \to \infty} P(A_n) =$

$\lim_{n \to \infty} F(b - 1/n) = F(b^-)$. $\square$

Example: A wire three feet long will be cut at random to form two pieces.

By "at random" we shall mean that within the interval $[0, 3]$, the probability of the cut falling within an interval I is proportional to the length of the interval (e.g. $P((0, 2)) = 2 \cdot P((1, 2))$.   Under the assumption that such a concept of probability does exist on $\mathscr{B}(\mathbf{R})$, we have that for each $x \in \mathbf{R}$, $P((-\infty, x])$ is proportional to the length of $[0, 3] \cap (-\infty, x]$.   Since $P([0, 3]) = 1$, we must have that the constant of proportionality is $1/3$.   Thus $F(x) = P((-\infty, x]) = 1/3(\text{length of } (-\infty, x] \cap [0, 3])$ so

$$F(x) = \begin{cases} 0 & \text{if } x \le 0 \\ x/3 & \text{if } 0 \le x \le 3 \\ 1 & \text{if } x \ge 3 \end{cases}$$

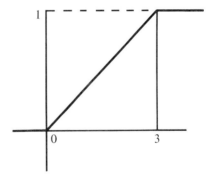

F satisfies all of the properties of Proposition 1 so there does exist a probability measure on $\mathscr{B}(\mathbf{R})$ whose distribution function is F.

In general, the probability of a set A can be interpreted in a sense as the proportion of $[0, 3]$ that A contains.

Example: Let $\mathbf{Q} = \{r_1, r_2, \ldots, r_n, \ldots\}$ and assign weights

$$w(t) = \begin{cases} 1/2^n & \text{if } t = r_n \text{ for some } n \\ 0 & \text{if not} \end{cases}.$$

Now w is a weight function since Q is countable and $\sum\limits_{t \in R} w(t) = \sum\limits_{1}^{\infty} 1/2^n = 1$. Thus by the treatment in the preceding section, w generates a probability measure on any σ-algebra of subsets of R and thus on $\mathscr{B}(R)$. This probability measure in turn generates a distribution function $F(x) = P((-\infty, x]) = \sum\limits_{t \in (-\infty, x]} w(t) = \sum\limits_{t \leq x} w(t)$. It is impossible to draw the graph of F since the particular ordering of Q is not given. We do know, however, several facts about F (namely (a)–(d) of Theorem 3) simply because it is a distribution function. Incidentally, it can be shown fairly easily that F is continuous at each irrational number and discontinuous at each rational!

Exercises:

1. Show that the set of irrational numbers is a Borel set in R.

2. Prove Theorem 2.

3. In the proof of part (b) of Proposition 1 it is proven only that $\lim\limits_{n \to \infty} F(-n) = 0$. Why does it follow that $\lim\limits_{a \to -\infty} F(a) = 0$?

4. Prove parts (c)–(h) of Theorem 4.

5. The interval $[0, 1]$ is cut once at random. Find the probability that the longest piece is no longer than $2/3$.

6. Using the probability measure involved in the preceding problem, find

   (a) $P(\{1/2\})$
   (b) $P(\{a\})$ for each $a \in R$
   (c) $P(Q)$.

7. A point is chosen at random inside a right triangle ABC (where the right angle is at A). What is the probability that the point is closer to A than to either B or C? Here "random" means that the probability of the point being inside a particular region within the triangle is proportional to the area of that region.

8. Accept the challenge implicit in the last sentence of this section.

3

Counting

In this chapter we shall develop several counting formulae that will help us compute probabilities.

Definition 1:   Let A and B be sets; then we define the *Cartesian product* of A and B by

$$A \times B = \{(a, b) \mid a \in A, b \in B\}$$

where $(a, b) = (c, d)$ only if $a = c$ and $b = d$.   The members of $A \times B$ are called *ordered pairs*.

For example, if $A = \{a_1, a_2, a_3\}$ and $B = \{b_1, b_2\}$, then $A \times B = \{(a_1, b_1), (a_2, b_1), (a_3, b_1), (a_1, b_2), (a_2, b_2), (a_3, b_2)\}$.   We can represent $A \times B$ in this case as points in a two-dimensional array as in figure 3.1.

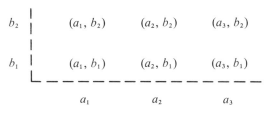

Figure 3.1

Proposition 1:   *If A and B are finite sets then*

$$\#(A \times B) = \#(A) \cdot \#(B).$$

Proof: Let $\#(A) = n$ and $\#(B) = m$; if $n = 0$ or $m = 0$ then $A \times B = \varnothing$ so $\#(A \times B) = 0 = n \cdot m = \#(A) \cdot \#(B)$. If $n \neq 0 \neq m$, then for each $a \in A$, there are m members of $A \times B$ which contain a as a first element. Thus in $A \times B$, there are n sets of m elements each, so $\#(A \times B) = n \cdot m = \#(A) \cdot \#(B)$. $\square$

The same method of counting allows us to prove the following principle. We state it in somewhat "non-mathematical" terms.

FUNDAMENTAL PRINCIPLE OF COUNTING: If one procedure has n possibilities for outcomes and after any one of these possibilities has occurred, another procedure has m possibilities, then the number of possible outcomes for the first procedure followed by the second is $n \cdot m$.

Proof: We may consider the collection of possible outcomes of the first procedure followed by the second as made up of elements of the form (a, b) where a is an outcome of the first procedure and b is an outcome of the second, given that a has occurred. For each of the n a's, there are m b's; this gives a total of $n \cdot m$ outcomes for the succession of procedures. $\square$

Example: How may two-digit numbers can be formed from the digits $\{1, 2, 3, 4, 5\}$?

Here our two procedures are "picking the first digit" and "picking the second digit." There are five ways to pick the first digit, but before we can tell how many ways we can choose the second digit, we must decide how to interpret the problem: do we allow repetitions, such as 33, or not? If repetitions *are* allowed, then there are five ways to choose the second digit so the total number of two-digit numbers is $5 \cdot 5 = 25$. If repetitions are not allowed, then there are only four ways to choose the second digit and the total number of two digit numbers is $5 \cdot 4 = 20$.

This counting method may be extended to more than two procedures as in the following example.

Example: In how many ways can a president, vice president, secretary, and treasurer be chosen from a club of 100 members if no person can hold more than one office?

There are 100 ways in which the president may be chosen and after that, 99 ways for the vice president. Let us consider the choosing of president and vice president as a single procedure (for which there are $100 \cdot 99$ possible outcomes). The secretary may then be chosen in 98 ways so there are $(100 \cdot 99) \cdot 98$ ways in which the first three offices may be filled. We combine the selection of president, vice president, and secretary into one procedure (for which there are $(100 \cdot 99) \cdot 98$ outcomes) and follow it by the selection of the treasurer, for which there are 97 possibilities. Thus the four offices may be filled in $100 \cdot 99 \cdot 98 \cdot 97$ different ways.

A computational and notational aid useful in treating this type of counting problem is the concept of the factorial.

Definition 2: We define the concept of the *factorial* of a non-negative integer n (denoted $n!$ and read "n factorial") inductively as follows:

(a) $0! = 1$,

(b) if $n \in \mathbf{N}$, then $n! = n \cdot (n - 1)!$.

For example, $1! = 1 \cdot 0! = 1 \cdot 1 = 1$

$2! = 2 \cdot 1! = 2 \cdot 1 = 2$

$3! = 3 \cdot 2! = 3 \cdot 2 \cdot 1 = 6$

$4! = 4 \cdot 3! = 4 \cdot 3 \cdot 2 \cdot 1 = 24$, etc.

It may seem somewhat odd to define $0!$ as 1 but this definition often allows us to state and prove formulae without considering several special cases. One such formula is that in Theorem 1 below.

Example: In how many ways can nine boys be assigned positions on a baseball team if there are only two boys who can be pitcher?

We have two procedures involved—picking the pitcher and then assigning the other positions.   There are two ways in which the pitcher may be picked and then there are eight boys for the other eight positions, thus those positions may be assigned in

$$8 \cdot 7 \cdot 6 \cdot 5 \cdot 4 \cdot 3 \cdot 2 \cdot 1 = 8!$$

ways.   The entire team may then be assigned in $2 \cdot (8!)$ ways.

Definition 3:   Let A be a set.   We define A^n inductively as follows:

(1) $A^1 = A$

(2)   if $n \in N$ and $n > 1$, then $A^n = A \times (A^{n-1})$.

There is a natural one-to-one correspondence between $A \times (A \times A)$ and $(A \times A) \times A$ (namely $(a, (b, c)) \leftrightarrow ((a, b), c))$ so we shall simply identify the two and denote them by $A \times A \times A$.   An element of $A \times A \times A$ is then of the form (a_1, a_2, a_3) where $a_1 \in A$, $a_2 \in A$, and $a_3 \in A$.   We treat other powers of A similarly and so we have $A^1 = A$,

$$A^2 = A \times A, A^3 = A \times A \times A, \text{etc. In general } A^n = \overbrace{A \times A \times \cdots \times A}^{n \text{ times}}$$

and the members of A^n are of the form $(a_1, a_2, \ldots, a_n)$ where each $a_i \in A$. The members of A^n are called ordered n-tuples and we note that $(a_1, a_2, \ldots, a_n) = (b_1, b_2, \ldots, b_n)$ only when $a_i = b_i$ for each integer $i = 1, 2, \ldots, n$.

Definition 4:   Let A be a finite, non-empty set with n elements.   A *permutation of A* is any member $(a_1, \ldots, a_n)$ of A^n with the property that $a_i \neq a_j$ if $i \neq j$.   A *permutation of m elements of A* is any member $(a_1, \ldots, a_m)$ of A^m with the property that $a_i \neq a_j$ if $i \neq j$.

A permutation of a set A may be considered as an *arrangement* of all members in that set while a permutation of m elements is an arrangement of some m of them.   We use the symbol "$P(n, m)$" to denote the number of permutations of m objects from a set of n.   It is clear that if $m > n$, then $P(n, m) = 0$; however, as a convention we define $P(n, m) = 1$ if $m = 0$ (the non-existent arrangement?).   The rationale behind this definition is the same as that behind the definition of $0!$ as 1: it allows us to state formulae without special cases (see, for example, Theorem 3).

Example: Compute $P(5, 2)$.

We must count the number of ways in which we can choose and arrange two elements from a set of five (with no repetitions). There are $5 \cdot 4 = 20$ ways in which this may be done so $P(5, 2) = 20$.

THEOREM 1: If $n \in \mathbb{N}$ and $m \in \mathbb{N}$, then

$$P(n, m) = \begin{cases} n \cdot (n - 1) \cdots (n - m + 1) = \dfrac{n!}{(n - m)!} & \text{if } m \leq n \\ 0 & \text{if } m > n \end{cases}.$$

Proof: We have already observed that if $m > n$, then $P(n, m) = 0$. For $m \leq n$, two proofs will be given. The first is a direct application of the Fundamental Principle of Counting and the second illustrates a method that will be used several times in this section: the concept of one object "generating" other objects.

Proof 1: There are n ways to pick the first element, $(n - 1)$ ways for the second, and so on down to the mth element for which there exist $(n - m + 1)$ ways. Thus

$$P(n, m) = \overbrace{n \cdot (n - 1) \cdot (n - 2) \cdots (n - m + 1)}^{m \text{ terms}}$$

$$= \frac{n(n - 1) \cdots (n - m + 1)(n - m)(n - m - 1) \cdots 3 \cdot 2 \cdot 1}{(n - m)(n - m - 1) \cdots 3 \cdot 2 \cdot 1}$$

$$= \frac{n!}{(n - m)!}.$$

Proof 2: As in the method of Proof 1, we establish that for each $k \in \mathbb{N}$, $P(k, k) = k!$. Now a permutation of all n elements can be chosen by picking the first m elements and then arranging the remaining $(n - m)$ elements. Thus for each permutation of m elements from the set of n, there are $(n - m)! = P(n - m, n - m)$ permutations of all n elements. Therefore

$$n! = P(n, n) = P(n, m) \cdot P(n - m, n - m)$$

$$= P(n, m) \cdot ((n - m)!).$$

The required formula now follows easily. $\square$

Example: A man is dealt four cards from a well-shuffled standard deck of fifty-two. What is the probability of his getting all four aces?

There are $P(4, 4) = 4!$ ways in which he could have been dealt the four aces. The total number of four-card hands that he could have been dealt is

$$P(52, 4) = \frac{52!}{(52 - 4)!} = \frac{52!}{48!}.$$

If each of these hands has the same probability, then the probability of (the event:) getting four aces is $4!/(52!/48!) = 4!\,48!/52!$ which is $1/270{,}725$.

Occasionally, the elements which we wish to arrange or permute may be indistinguishable. For example let us consider the number of permutations of four flags, of which three are blue and the other red. Now our formula gives $P(4, 4) = 4! = 24$ permutations in all. However, to the eye, some of these are the same; if the set consists of $\{R, B_1, B_2, B_3\}$ then the permutation (R, B_1, B_2, B_3) looks the same as (R, B_3, B_1, B_2).

THEOREM 2:   Let A be a set with n elements and suppose that

$$n_1 \text{ of them are alike of one kind}$$

$$n_2 \text{ of them are alike of another}$$

$$\vdots$$

$$n_k \text{ of them are alike of another.}$$

Then the number of distinguishable permutations of all n elements is

$$\frac{n!}{n_1!\, n_2! \cdots n_k!}.$$

Proof: Let m be the number of distinguishable permutations. If none of the elements were alike there would be $n!$ permutations in all. Now each of the m distinguishable permutations "generates" several other permutations which are indistinguishable from it. In particular, there are $n_1!$ arrangements of those of the first type, $n_2!$ arrangements of those of the second type, and so on. Thus each distinguishable permutation generates $n_1!\, n_2! \cdots n_k!$ permutations that are indistinguishable from it, hence there are $m \cdot (n_1! \, n_2! \cdots n_k!)$ permutations generated in this fashion. But all $n!$ permutations must be generated in this way so

$$n! = m(n_1! \, n_2! \cdots n_k!)$$

from which the formula follows. □

Example: How many possible arrangements are there of all the letters in the word "TENNESSEE"?

Here there are nine letters in all, but several repetitions, namely, four E's, two N's, and two S's, so the total number of arrangements is

$$\frac{9!}{4! \cdot 2! \cdot 2!} = \frac{9 \cdot 8 \cdot 7 \cdot 6 \cdot 5}{2 \cdot 2} = 3780.$$

Notice that we could also have considered the one T and we would have gotten

$$\frac{9!}{4! \cdot 2! \cdot 2! \cdot 1!}$$

which is the same answer since $1! = 1$.

In the preceding paragraphs, we have been concerned with gathering together m elements from a set of n into an arrangement; often, however, we may only be interested in the elements involved in the arrangement. In the case of the four-card hand investigated above, the order in which the cards were dealt was unimportant. We are only interested in a subset of four cards from a set of fifty-two. These subsets are sometimes called combinations; we use $C(n, m)$ to denote the number of subsets (combinations) of m elements which may be chosen from a set of n.

Example: Compute $C(4, 2)$

Let $A = \{a, b, c, d\}$, then the subsets consisting of two elements of A are

$$\{a, b\}, \{a, c\}, \{a, d\}$$

$$\{b, c\} \quad \{b, d\}$$

$$\{c, d\}.$$

Thus $C(4, 2) = 6$.

Note that $\{b, a\}$ was not included in the above list because this is the same subset as $\{a, b\}$; for subsets or combinations, order is not important.

THEOREM 3:   If n and m are non-negative integers, then

$$C(n, m) = \begin{cases} \dfrac{P(n, m)}{m!} = \dfrac{n!}{m! \, (n - m)!} & \text{if } 0 \le m \le n \\ 0 & \text{if } m > n. \end{cases}$$

Proof: It is clear that if $m > n$, then $C(n, m) = 0$. If $0 \le m \le n$, then each subset of m elements generates $m!$ permutations by simply rearranging the elements; all permutations are generated in this fashion. Thus the total number of permutations of m elements from a set of n is $P(n, m) = C(n, m) \cdot m!$ from which the formula follows directly. $\square$

Example: How many committees of five members may be appointed from a club of fifteen members if the first committee member chosen is to be chairman?

We may split this problem into two parts—choosing a chairman, and then choosing the other members. Now the chairman may be chosen in fifteen ways. After he is chosen, the order of the selection of the other four members is unimportant so there are $C(14, 4)$ ways in which they may be chosen. Thus the committee may be appointed in $15 \cdot C(14, 4) = 15(14!/(10! \, 4!)) = 15{,}015$ ways.

Example: Of the 100 members of the legislature of a certain state, there are 75 Democrats and 25 Republicans. In how many ways can a committee of 10 be chosen so that 7 Democrats and 3 Republicans are included?

In this example, there are two procedures involved: picking seven Democrats and then picking three Republicans. The first can be done in $C(75, 7) = 75!/(7! \, 68!)$ ways and the second $C(25, 3) = 25!/(3! \, 22!)$. Thus the entire committee can be chosen in

$$C(75, 7) \cdot C(25, 3) = \frac{75! \, 25!}{3! \, 7! \, 22! \, 68!}$$

ways.

The numbers $C(n, m)$ also appear in the apparently unrelated result below.

BINOMIAL THEOREM: If r and s are real numbers and $n \in \mathbb{N}$, then

$$(r + s)^n = \sum_{k=0}^{n} C(n, k) r^k s^{n-k}.$$

Proof: The Binomial Theorem may be proved by induction (see exercise 4) but the proof does not show how the coefficients $C(n, k)$ arise, so we shall consider a different proof. When $(r + s)^n$ is "multiplied out" each term is of a form like

$$\overbrace{r \cdot r \cdot s \cdot r \cdot \ldots \cdot r.}^{n \text{ terms}}$$

If there are k r's, then the term can be expressed as $r^k s^{n-k}$. The problem of finding the coefficient of $r^k s^{n-k}$ in the statement of the theorem is just the problem of counting the terms yielding $r^k s^{n-k}$. This is simply the problem of computing the number of permutations of n objects of which k are alike (namely r's) and $n - k$ are alike (namely s's); this number is $n!/(k! \, (n - k)!) = C(n, k)$. $\square$

The procedure of selecting a subset B of a set A can be considered as a procedure for splitting up A—into B and $A \backslash B$. These sets are disjoint and together they compose A. This observation leads us to the following definition.

Definition 5:   Let A be a set and $\{B_\alpha : \alpha \in \mathfrak{A}\}$ a collection of subsets of A. $\{B_\alpha : \alpha \in \mathfrak{A}\}$ is a *partition* of A provided that

    (a)   if α and β are different members of $\mathfrak{A}$, then $B_\alpha \cap B_\beta = \varnothing$
    (b) $\bigcup \{B_\alpha : \alpha \in \mathfrak{A}\} = A$.

A member B_α of the partition is called a *cell*. The *ordered partition* associated with $\{B_\alpha : \alpha \in \mathfrak{A}\}$ is $\{(\alpha, B_\alpha) \mid \alpha \in \mathfrak{A}\}$.

A partition of a set is simply a splitting up of that set into disjoint subsets. We illustrate the concept of (and necessity for) ordered partitions by the following example.

Example: In a class of ten, a teacher will give out five passing grades and five failing grades.

We shall take the class of ten (as in the case of most universities, names will be replaced by numbers, say $\{1, 2, 3, 4, 5, 6, 7, 8, 9, 10\}$) and partition it into two groups, B_1 and B_2. The students in B_1

will receive passing grades and those in B_2, failing grades; for example $B_1 = \{1, 2, 3, 4, 5\}$, $B_2 = \{6, 7, 8, 9, 10\}$. Now the partition is

$$\{\{1, 2, 3, 4, 5\}, \{6, 7, 8, 9, 10\}\}$$

and this is exactly the same *collection* of sets as

$$\{\{6, 7, 8, 9, 10\}, \{1, 2, 3, 4, 5\}\}.$$

The latter is the collection formed by reversing all the grades. Now certainly the two partitions are the same but everyone should agree that some difference should exist; it is important to know not only the elements of the cells but also the indices involved. This is why, in an ordered partition, the index is included as part of the cell.

We may count the numbers of ordered partitions of a certain type in much the same way that we counted the number of subsets of a given size.

THEOREM 4:   Let A be a set with n elements ($n \in \mathbb{N}$). The number of possible partitions of A into k cells such that there are

> n_1 elements in the first cell,
>
> n_2 elements in the second cell,
>
> $\vdots$
>
> n_k elements in the kth cell

is $n!/(n_1!\, n_2! \cdots n_k!)$.

Proof: As before, we shall consider how an ordered partition can generate permutations. Suppose the cells are lined up with cell 1 followed by cell 2, etc. The n_1 elements of cell 1 can be arranged in $n_1!$ ways, those in cell 2, in $n_2!$ ways, and so on. Thus each ordered partition generates $n_1!\, n_2! \cdots n_k!$ permutations and each permutation is generated in this fashion from some ordered partition. If m is the number of ordered partitions of the type specified above, then

$$m \cdot (n_1!\, n_2! \cdots n_k!) = P(n, n) = n!. \quad \square$$

Example: In how many ways may four people be dealt thirteen cards each from a standard deck of fifty-two?

Since it does make a difference who gets a particular hand, then we are dealing with ordered partitions and the answer is

$$\frac{52!}{13! \, 13! \, 13! \, 13!}.$$

Example: A regular die is to be tossed ten times. In how many ways can there be obtained exactly five 2's and three 6's?

Let $A = \{a_1, a_2, \ldots, a_{10}\}$ and we have the problem of separating out five tosses (to be assigned the value 2) and three tosses (to be assigned the value 6); this leaves two tosses. Thus we wish to partition A into cells of 5, 3, and 2 elements. There is an order implicit here because there can be no confusion between a cell with, for example, 5 elements and a cell with 3. Thus the number of partitions is $10!/(5! \, 3! \, 2!)$ and for the remaining (non-5 and non-3) two cases, there are $4 \cdot 4 = 16$ possibilities. Therefore the answer is $16 \cdot 10!/(5! \, 3! \, 2!)$.

Example: In how many ways can a group of twenty boys be divided into four teams of five each?

Here we wish to partition a set of twenty elements into four cells of five each. Whether or not they should be ordered partitions depends on what will happen after they are divided. If the teams will play a single elimination tournament, then clearly the order of the teams is important and the answer is

$$\frac{20!}{5! \, 5! \, 5! \, 5!}.$$

If the teams will play a round robin tournament (each team plays each other team) then order is unimportant. Since each unordered partition may generate 4! ordered partitions, then the number of unordered partitions is $(1/4!)(20!/(5! \, 5! \, 5! \, 5!))$.

Example: In how many possible ways can ten people from a group of fifteen be seated around a round table?

If the people were seated in a row, then the answer would be $P(15,10)$. However in the case of a round table, there are some

arrangements that are essentially the same. For example, if we were
interested in seating three people, then the arrangements shown in
figure 3.2 are really the same. For ten people we can rotate nine

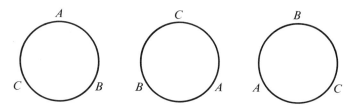

Figure 3.2

times to produce nine other arrangements equivalent to a given one.
Thus there are only 1/10 as many distinct arrangements so the correct
answer is $P(15,10)/10$.

Example: In how many possible ways can ten beads from a group of
fifteen be placed on a string to form a necklace?

This appears to be the same problem as that in the preceding example
but there is one large difference. In this case, each arrangement may
be flipped over to produce another. For example, with three, the
arrangements in figure 3.3 are equivalent. Thus there are only half
as many arrangements as before so the answer is $P(15, 10)/2 \cdot 10$.

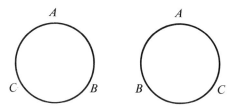

Figure 3.3

Example: A man is dealt four cards from a well-shuffled standard deck
of fifty-two. What is the probability of his getting all four aces?

The total number of ways in which he could be dealt a hand of four
cards is $C(52, 4) = 52!/4! \, 48!$ and only one of these has all four
aces. If all hands have the same probability, then the probability of
getting four aces is $1/C(52, 4) = 4! \, 48!/52! = 1/270,725$.

Note that the example preceding is exactly the same as the first example after Theorem 1. In the first case we found the answer by using permutations and in the second, by using combinations. Both methods gave the same answer—the first used a particular probability measure space and the second, another; both are reasonable models for the "real-life" process.

Exercises:

1. Find the total number of ways in which five persons can be seated in a row of eight chairs. If the seating occurs at random, what is the probability that the two end seats will be left unoccupied?

2. In how many ways can all of the letters in the word "MISSISSIPPI" be arranged?

3. Prove that for each $n \in N$ and $r \in N$,

$$C(n + 1, r) = C(n, r - 1) + C(n, r).$$

This proves that the binomial coefficients can be generated by Pascal's Triangle (figure 3.4), where 1's are placed along the sides and each other number is obtained by summing the two immediately above it (e.g. $4 = 1 + 3$ in the fifth line).

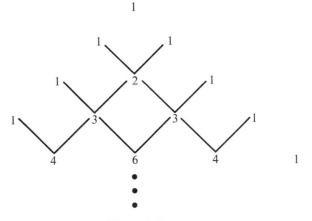

Figure 3.4

4. Use the relationship in exercise 3 to prove the Binomial Theorem by induction.

5. Prove that for each $n \in N$, $\sum\limits_{k=0}^{n} C(n, k) = 2^n$.

6. Prove that if A is a set with n elements ($n \in$ N), then $\mathscr{P}(A)$ has 2^n members.

7. Prove that if $n \in$ N, and r, s, and t are real numbers then

$$(r + s + t)^n = \sum_{\substack{i+j+k=n \\ 0 \le i,j,k \le n}} \left(\frac{n!}{i! \, j! \, k!} \right) r^i s^j t^k.$$

(Hint: Use a method similar to that used in the proof of the Binomial Theorem.)

8. In a certain academic department of a university, there are forty-five graduate students of whom three are female. If the students are assigned to twenty-two offices there will be twenty-one two-man offices and one three-man office. If office assignments are made at random, what is the probability that the three females will be placed in the three-man office?

9. In how many possible ways can twenty students be assigned grades of A, B, C, D, and F if there are to be two A's, three B's, ten C's, three D's, and two F's?

10. A class contains nine boys and seven girls. If a committee of ten is chosen at random, what is the probability that there are seven boys and three girls on the committee?

11. In how many ways can positions on a baseball team be assigned to nine players if one player can play only shortstop or third base?

12. Each of ten men has a well-shuffled deck of cards and will draw a card from his deck. What is the probability that no two pick the same card? What is the probability that at least two pick the same card?

13. An urn contains three white balls and five black balls. Two balls are drawn from the urn in succession (without replacement). What is the probability of getting two white balls?

14. What is the probability that REED is spelled if the letters of the word "DEER" are arranged at random?

15. A box contains forty good and ten defective fuses. Ten fuses are selected at random from the box. What is the probability that they are all good?

16. Four red flags and four black flags are arranged at random in a line. What is the probability that they alternate?

17. A three-digit number is formed at random from the digits 1, 2, 3, 4, 5, 6 (no digit may be used more than once). What is the probability that it is even? What is the probability that it is less than 378? What is the probability that it is either even or less than 378?

18. Four boys and four girls sit at random in a row of eight chairs. What is the probability that the girls sit together? What is the probability that they do not?

19. At a political meeting, there are three factions represented, each by four persons. In how many ways can the twelve politicians be seated in a row of twelve chairs if members of the various factions insist on sitting together?

20. Modify the preceding problem by having the politicians sit at a round table.

21. In a strand of the DNA molecule, there exist four different types of *bases* denoted by A, T, G, and C. The genetic code for the DNA is read by taking sequences of three bases (such as ATA or GAA). How many possible such sequences exist?

4

Conditional Probability
and Independence

4.1 Conditional Probability

In computing the probability of a particular event, there is sometimes the possibility of using additional information. For example, if a fair die is thrown, then $P(3) = 1/6$. However, if we know the additional fact that the outcome is even, then a 3 is impossible so the probability is reduced to zero. If we know the additional fact that the outcome is odd, then the probability of a 3 is raised to 1/3. We have adjusted the probability of getting a 3 by computing the amount of the given event (odd or even) that {3} accounts for. We make this concept rigorous in the following definition.

Definition 1: Let (S, Σ, P) be a p.m.s. and $E \in \Sigma$ such that $P(E) \neq 0$. For each $A \in \Sigma$, we define "the conditional probability of A, given E" by

$$P(A \mid E) = \frac{P(A \cap E)}{P(E)} .$$

It is relatively easy to prove that as a function in A, $P(A \mid E)$ *is* a probability measure on (S, Σ) (see exercises, page 21).[1] As we have seen, $P(A \mid E)$ may be greater than $P(A)$ or it may be less than $P(A)$; there is also the possibility that $P(A \mid E) = P(A)$. Another form of the equation in Definition 1 is important also: this is $P(A \cap E) = P(E) \cdot P(A \mid E)$.

One concept that is useful in dealing with conditional probability is that of "tree diagrams." For the example at the beginning of this section, we have the diagram in figure 4.1. At the first level of the tree, we have divided the outcomes into odd (O) and even (E) and placed the appropriate probabilities on the branches leading to these events. At the second level, we have again divided the outcomes, this time into "$\{3\}$" and "not $\{3\}$" and placed the appropriate *conditional* probabilities on the branches leading to these events.

We could have also considered a tree in which $\{3\}$ was omitted from the E branch since $P(\{3\} \mid E) = 0$. Just as there may exist many probability measure spaces which provide a model for a particular process, there may

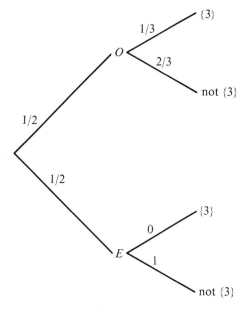

Figure 4.1

[1] This function may be denoted by $P((\cdot)|E)$ where $(\cdot)$ indicates the position of the variable.

exist several trees which may be appropriate. For example, we could have considered the trees in figure 4.2.

At each branch point, or node (*O* and *E* are nodes), it *is* necessary to have the probabilities leading from it to sum to 1. Also, no element should be included in more than one branch leading from a particular node. That is, the division at each node should be a partition of the possible outcomes.

Example: Three dice are located in a box; two of them fair, regular dice and the other fair, but with two 1's, two 3's, and two 5's. A die is selected at random and thrown.

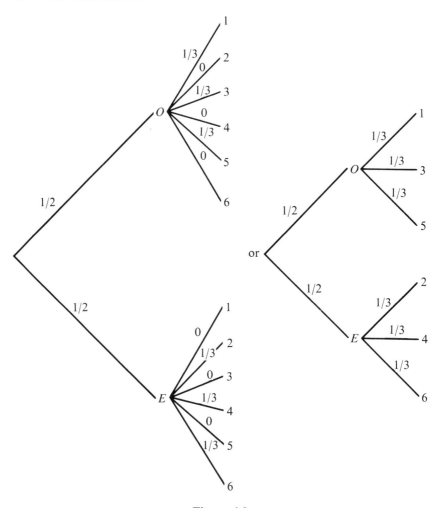

Figure 4.2

Here, one possible tree is shown in figure 4.3, where R represents "regular" and N represents "not regular." Now we may compute such probability as $P(\{3\})$ as follows

$$P(\{3\}) = P((\{3\} \cap R) \cup (\{3\} \cap N))$$

$$= P(\{3\} \cap R) + P(\{3\} \cap N)$$

$$= P(\{3\} \mid R) \cdot P(R) + P(\{3\} \mid N) \cdot P(N),$$

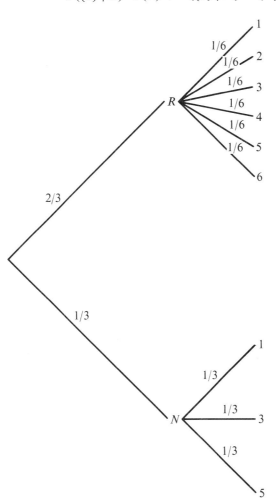

Figure 4.3

by Definition 1 of this section. Each of these products is obtained by multiplying the probabilities along appropriate branches leading to "3"; thus $P(\{3\} = (2/3) \cdot (1/6) + (1/3) \cdot (1/3) = 2/9$. Similarly, we find that $P(\{4\}) = (2/3) \cdot (1/6) + (1/3) \cdot 0 = 1/9$.

We could also ask another type of question in the above example. Suppose a die is picked at random and tossed and a three is obtained; what is the probability that the die thrown was regular? Here we are asking for

$$P(R \mid \{3\}) = \frac{P(R \cap \{3\})}{P(\{3\})} = \frac{1/9}{2/9} = \frac{1}{2}.$$

An important tool in answering such ex post facto questions is the following.

BAYES' THEOREM: Let (S, Σ, P) be a p.m.s. and $\{A_1, A_2, \ldots, A_n\}$ a partition of S with $A_i \in \Sigma$ and $P(A_i) > 0$ for $i = 1, 2, \ldots, n$. Let $B \in \Sigma$ such that $P(B) > 0$; then for each i $(1 \leq i \leq n)$, we have

$$P(A_i \mid B) = \frac{P(A_i) \cdot P(B \mid A_i)}{\displaystyle\sum_{j=1}^{n} [P(A_j) \cdot P(B \mid A_j)]}.$$

Proof: Note from the definition of conditional probability that if E and F are members of Σ and $P(F) > 0$, then $P(E \cap F) = P(F) \cdot P(E \mid F)$.

$$P(B) = P(S \cap B) = P\left(\left(\bigcup_1^n A_j\right) \cap B\right) = P\left(\bigcup_1^n (A_j \cap B)\right)$$

$$= \sum_{j=1}^{n} P(A_j \cap B).$$

Thus

$$P(A_i \mid B) = \frac{P(A_i \cap B)}{P(B)} = \frac{P(A_i \cap B)}{\displaystyle\sum_{j=1}^{n} P(A_j \cap B)} = \frac{P(A_i) \cdot P(B \mid A_i)}{\displaystyle\sum_{j=1}^{n} [P(A_j) \cdot P(B \mid A_j)]}. \quad \square$$

Bayes' Theorem is simply a restatement of the definition of conditional probability in a form that can be used easily with a tree diagram, as in figure 4.4.

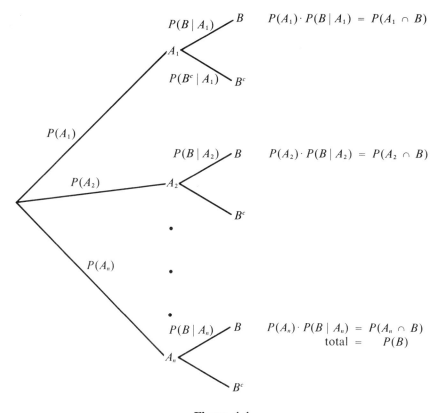

$P(A_1) \cdot P(B \mid A_1) = P(A_1 \cap B)$

$P(A_2) \cdot P(B \mid A_2) = P(A_2 \cap B)$

$P(A_n) \cdot P(B \mid A_n) = P(A_n \cap B)$
total $= P(B)$

Figure 4.4

Example: Suppose your roommate has a math quiz today and figures that the probability of his passing is 3/4. If he passes, then the probability that he will go out drinking to celebrate is 8/10; whereas, if he fails, then the probability that he will go out to drown his sorrows is 1/2. Later tonight you go to your room and find him passed out, reeking of alcohol, and with a smile on his face. What is the probability that the smile is a result of his having passed the quiz?

Let P represent "pass" and F represent "fail"; let D represent "drinking" and N represent "not drinking." Then we are asking for $P(P \mid D)$. An appropriate tree would be figure 4.5. Thus

$$P(P \mid D) = \frac{6/10}{6/10 + 1/8} = \frac{6/10}{29/40} = \frac{24}{29}.$$

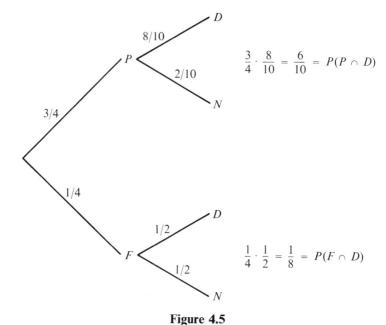

$$\frac{3}{4} \cdot \frac{8}{10} = \frac{6}{10} = P(P \cap D)$$

$$\frac{1}{4} \cdot \frac{1}{2} = \frac{1}{8} = P(F \cap D)$$

Figure 4.5

Exercises:

1. A pair of fair dice is tossed; find the probability that the sum of the numbers obtained is seven, given that
   (a) the outcome of the first die is divisible by three.
   (b) the outcome of at least one die is divisible by three.

2. In a particular high school, the foreign language teacher reports an enrollment of 20 percent of the school in French, 30 percent in Spanish, and notes that 5 percent of the school is enrolled for both. A student is picked at random from the school and it is found that he is enrolled for Spanish. What is the probability that he is *not* enrolled for French?

3. In the secretarial pool of a certain company, there are three secretaries. Secretary A types 50 percent of all letters but 40 percent of her letters have errors. Secretary B types 30 percent of all letters and 10 percent of her letters have errors. Secretary C types the other letters and only 5 percent of her letters have errors. A letter is sent to the pool and comes back with no errors. What is the probability that it was typed by secretary A, by secretary B, by secretary C?

4. In a certain country, 10 percent of the inhabitants are infected with a particular disease. A new diagnostic test is devised which has the property that 95 percent of subjects who have the disease test out positive. However, 5 percent of subjects who do not have the disease also test out positive. A person is selected at random from the population of the country and is given the test. The results are positive; what is the probability that he is infected with the disease?

4.2 Independence

In computing the conditional probability of an event A, given an event B, it may happen that $P(A \mid B) = P(A)$. For example, in the toss of a fair die, the probability that the outcome is divisible by 3 is $1/3$. Suppose the outcome is even, then

$$P(\text{divisible by 3} \mid \text{even}) = \frac{1/6}{1/2} = \frac{1}{3}.$$

This is the type of situation we wish to investigate in this section. However, since $P(A \mid B)$ is defined only when $P(B) > 0$, we seek a statement equivalent to "$P(A \mid B) = P(A)$" when $P(B) > 0$ but which is valid for a wider class of events.

Proposition 1:   Let (S, Σ, P) be a p.m.s. and A and B be elements of Σ with $P(B) > 0$.   These are equivalent:

(a) $P(A \mid B) = P(A)$

(b) $P(A \cap B) = P(A) \cdot P(B)$.

Proof: $P(A \mid B) = P(A) \Leftrightarrow \dfrac{P(A \cap B)}{P(B)} = P(A)$

$$\Leftrightarrow P(A \cap B) = P(A) \cdot P(B). \quad \square$$

Note that (b) is symmetric in A and B, thus if $P(A) > 0$ and $P(B) > 0$, then

$$P(A \mid B) = P(A) \Leftrightarrow P(B \mid A) = P(B).$$

Definition 1:   Let (S, Σ, P) be a p.m.s. and A and B be members of Σ. We shall say that A and B form an *independent pair* (or simply, A and B are independent) provided

$$P(A \cap B) = P(A) \cdot P(B).$$

If A is any event, then A and $\varnothing$ are always independent and A and S are always independent (see exercise 8, page 62. The following result is left as an exercise.)

Proposition 2:   *Let (S, Σ, P) be a p.m.s. and A, $B \in \Sigma$ such that A and B are independent. Then*

(a) $A^c = S \backslash A$ and B are independent,

(b) A^C and B^C are independent.

If A and C are independent and B and C are independent, then it does not necessarily follow that $(A \cup B)$ and C are independent (see exercise 3, page 61).

Students sometimes express independence informally by saying two events A and B are independent if they "don't have anything to do with one another." This in turn is sometimes interpreted as "$A \cap B = \varnothing$", however, this is rarely the correct interpretation. $A \cap B = \varnothing$, and A and B independent imply that $0 = P(A \cap B) = P(A) \cdot P(B)$ so $P(A) = 0$ or $P(B) = 0$. What should be meant is that knowing the occurrence of one does not affect the probability of the other.

The concept of independence motivates one model for a succession of happenings. If (S_1, Σ_1, P_1) and (S_2, Σ_2, P_2) are probability measure spaces, then $S_1 \times S_2$ is the set of all possible pairs of outcomes, one from S_1 followed by one from S_2. We seek a reasonable probability measure on (some) subsets of $S_1 \times S_2$ such that, in some sense, events in S_1 are independent of events in S_2. The situation is simplest in the discrete case.

THEOREM 1:   Let (S_1, Σ_1, P_1) and (S_2, Σ_2, P_2) be discrete probability measure spaces with weight functions w_1 and w_2, respectively.

(1)   The function $w \colon S_1 \times S_2 \to \mathsf{R}$ by $w(a, b) = w_1(a) \cdot w_2(b)$ is a weight function (and thus generates a probability measure P on $\mathscr{P}(S_1 \times S_2)$).

(2)   If $A \in \Sigma_1$ and $B \in \Sigma_2$, then

$$P(A \times B) = P_1(A) \cdot P_2(B).$$

Proof: (1) Since w_1 and w_2 vanish outside countable sets, so does w (why?). Now

$$\sum_{(a,b) \in S_1 \times S_2} w(a, b) = \sum_{a \in S_1} \sum_{b \in S_2} w_1(a) w_2(b)$$

$$= \sum_{a \in S_1} w_1(a) \left(\sum_{b \in S_2} w_2(b) \right)$$

$$= \sum_{a \in S_1} w_1(a)(1)$$

$$= 1.$$

(2) $P(A \times B) = \sum_{(a,b) \in A \times B} w(a, b) = \left(\sum_{a \in A} w_1(a) \right) \left(\sum_{b \in B} w_2(b) \right)$

$$= P_1(A) \cdot P_2(B). \quad \square$$

The independence condition which we sought is satisfied since $A \in \Sigma_1$ corresponds to $A \times S_2$, $B \in \Sigma_2$ corresponds to $S_1 \times B$; part (2) of the theorem gives

$$P((A \times S_2) \cap (B \times S_1)) = P(A \times B) = P_1(A) \cdot P_2(B)$$

$$= P(A \times S_2) \cdot P(S_1 \times B).$$

Example: A coin is weighted so that $P(H) = 1/3$; the coin is tossed twice. What is the probability of obtaining two heads?

The individual weight functions referred to in the preceding theorem are

$$w_i(H) = 1/3, \, w_i(T) = 2/3 \quad \text{for} \quad i = 1, 2,$$

the sample spaces, of course, are $S_1 = S_2 = \{H, T\}$. For the succession of tosses, the appropriate sample space is $S_1 \times S_2 = \{(H, H), (H, T), (T, H), (T, T)\}$ and the weight function is

$$w(H, H) = (1/3) \cdot (1/3) = 1/9$$

$$w(H, T) = w(T, H) = (1/3) \cdot (2/3) = 2/9$$

$$w(T, T) = (2/3) \cdot (2/3) = 4/9.$$

Thus we find that $P(\text{two heads}) = 1/9$.

In the case of a non-discrete p.m.s., the situation is much more difficult. Here $\mathcal{M} = \{A \times B \mid A \in \Sigma_1, B \in \Sigma_2\}$ forms a collection known as a semi-algebra (see exercise 9, page 14, for a related example of a semi-ring). The function $P: \mathcal{M} \to \mathsf{R}$ by

$$P(A \times B) = P(A) \cdot P(B)$$

has properties which are strong enough to ensure that it can be extended to a unique probability measure on the smallest σ-algebra containing $\mathcal{M}$. For a treatment of this semi-algebra to σ-algebra extension, the reader is referred to *Real Analysis* by H. L. Royden, published by Macmillan, 1963.

Example: A point will be chosen at random from the square whose vertices are $(0, 0)$, $(0, 1)$, $(1, 0)$, and $(1, 1)$.

By "at random" we shall mean that the probability of the chosen point occurring inside a given rectangle inside the square is proportional to the area of that rectangle. We shall treat the choice of a two-dimensional point as a succession of choices: the choice of the first coordinate followed by the choice of the second, and we shall treat the choices as independent. Each coordinate must be from the interval $[0, 1]$ and it is easy to see that the probability of a coordinate being within a subinterval of $[0, 1]$ is proportional to the length of that subinterval; the constant of proportionality must be 1 since the length of $[0, 1]$ is 1. The induced probability measure on $\mathcal{B}(\mathsf{R})$ gives the distribution function in figure 4.6.

Let P_1 be the induced probability measure on $\mathcal{B}(\mathsf{R})$, then by our previous discussion the function $P: \mathcal{M} \to \mathsf{R}$ (where $\mathcal{M} = \{A \times B \mid A, B \in \mathcal{B}(\mathsf{R})\}$) by $P(A \times B) = P_1(A) \cdot P_1(B)$ can be

$$F(x) = \begin{cases} 0 & \text{if } x < 0 \\ x & \text{if } 0 \leq x < 1 \\ 1 & \text{if } x \geq 1 \end{cases}$$

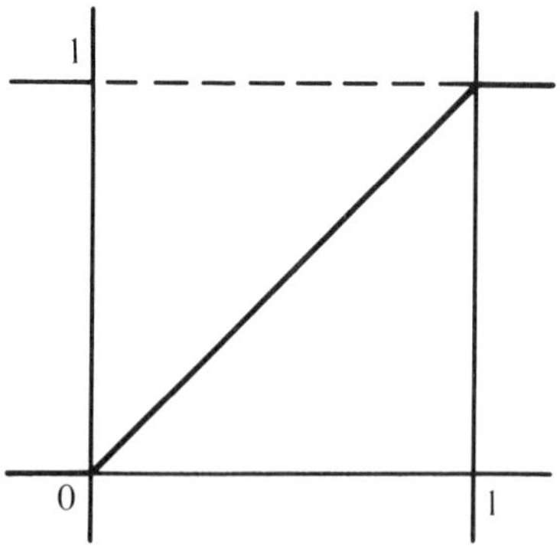

Figure 4.6

extended to a unique probability measure on the smallest σ-algebra containing $\mathcal{M}$. In particular, if $R = [a, b] \times [c, d]$ is a rectangle contained in the square $[0, 1] \times [0, 1]$, then $P(R) = P_1([a, b]) \cdot P_1([c, d]) = (b - a) \cdot (c - d) =$ area of R. This is the probability measure required in the problem. It can be shown (with a fair amount of difficulty) that for most common subsets A of $[0, 1] \times [0, 1]$, $P(A) = $ area (A). In particular, we may compute the probability that the point chosen be inside a circle with radius 1 and centered at $(0, 0)$ by simply finding the area involved, namely $(1/4)\pi(1^2) = \pi/4$.

Example: A point is chosen at random inside a circle of radius 1. What is the probability that it is closer to the center than to the rim?

Solution I: We shall pick the rectangular coordinates of the point independently. First pick a point in the square $[-1, 1] \times [-1, 1]$ such that regions of equal area have equal probability. Then the probability of being in a region is proportional to the area of that region (see the preceding example for details). Since the area of $[-1, 1] \times [-1, 1]$ is 4, then for each region A, $P(A) = (1/4) \cdot$ area(A). Let C be the region inside the circle of radius 1 centered at $(0, 0)$ and A be the region inside the circle of radius $1/2$ centered at $(0, 0)$. We simply require that the point be in C and then find the conditional probability that it be in A

$$P(A \mid C) = \frac{P(A \cap C)}{P(C)} = \frac{P(A)}{P(C)} = \frac{(1/4)(\pi(1/2)^2)}{(1/4)(\pi(1)^2)} = \frac{1}{4}.$$

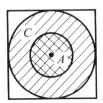

Figure 4.7

Solution II: We shall pick the polar coordinates of the point independently. If C is the interior of the circle centered at the origin with radius 1, then $C = \{(r, \theta): 0 \le r < 1, 0 \le \theta < 2\pi\}$ so it can be identified with $[0, 1] \times [0, 2\pi)$. As before, we pick r in $[0, 1)$ such that the probability of r being in an interval of I in $[0, 1)$ is proportional to the length of that interval. Since the length of $[0, 1)$ is 1, then $P_1(I) = \text{length}(I)$. In a similar fashion we choose θ in $[0, 2\pi)$ such that if I is an interval in $[0, 2\pi)$, then $P_2(I) = \text{length}(I) \cdot 1/2\pi$. Now the region A mentioned in Solution I is simply

$$\{(r, \theta): 0 \leq r < 1/2, 0 \leq \theta < 2\pi\}$$

which can be identified with $[0, 1/2) \times [0, 2\pi)$. Thus if r and θ are picked independently, then

$$P(A) = P_1([0, 1/2)) \cdot P_2([0, 2\pi)) = (1/2) \cdot 1 = 1/2.$$

Now which of the above "solutions" is correct? The answer is that either (or neither) can be correct provided the phrase "at random" in the statement of the problem is interpreted in the appropriate fashion. There are still other solutions to the problem depending on other interpretations of "at random." Each comes from a different *model* of the process of picking a point inside the circle.

We shall consider problems similar to that in the preceding example when we discuss jointly distributed random variables.

We extend our concept of independence to more than two events by the following definitions.

Definition 2:   Let (S, Σ, P) be a p.m.s. and A_1, A_2, and A_3 be events. $\{A_1, A_2, A_3\}$ is an independent collection provided that

(a)   each subcollection of two sets is independent,

(b) $P(A_1 \cap A_2 \cap A_3) = P(A_1) \cdot P(A_2) \cdot P(A_3)$.

Part (a) of this definition simply says that A_1 and A_2, A_1 and A_3, and A_2 and A_3 are independent. The following example and exercise 4 of this section show that there is no redundancy in listing both (a) and (b) above.

Example: Let $S = \{1, 2, 3, 4\}$, $\Sigma = \mathscr{P}(S)$, and let P be the equi-probable measure on Σ.

If $A_1 = \{1, 2\}$, $A_2 = \{2, 3\}$, and $A_3 = \{3, 1\}$, then it is easily seen that $\{A_1, A_2, A_3\}$ is pairwise independent (each pair is independent) but

$$P(A_1 \cap A_2 \cap A_3) = P(\emptyset) = 0 \neq 1/8 = (1/2) \cdot (1/2) \cdot (1/2)$$
$$= P(A_1) \cdot P(A_2) \cdot P(A_3).$$

Definition 3:   Let $\{A_i\}_1^\infty$ be a sequence of events in a p.m.s. (S, Σ, P).

(1) For each integer $m > 2$, the collection $\{A_1, A_2, \ldots, A_m\}$ of m sets is an independent collection provided that
(a)   each subcollection of $m - 1$ sets is independent
(b) $P(A_1 \cap A_2 \cap \cdots \cap A_m) = P(A_1) \cdot P(A_2) \cdot \ldots \cdot P(A_m)$.

(2) The collection $\{A_1, A_2, \ldots, A_m, \ldots\}$ is an independent collection provided that each finite subcollection is independent.

It is interesting to note that a condition similar to (b) in (1) follows for a countably infinite collection simply from (2) (see exercise 5 of this section). It is not hard to show that if $\{A_1, \ldots, A_m, \ldots\}$ is independent, then so is $\{B_1, \ldots, B_m, \ldots\}$ where for each i, $B_i = A_i$ or $B_i = (A_i)^c = S\backslash A_i$; this provides an extension of Proposition 2. We can prove a theorem similar to Theorem 1 for any finite collection of probability measure spaces by exactly the same methods that were used in Theorem 1.

Exercises:

1. A single fair die is tossed until a 3 is obtained. Let $S = \mathbb{N} = \{1, 2, 3, \ldots\}$ and $\Sigma = \mathscr{P}(S)$ and assume that tosses of the die are independent.
    (a)   Define explicitly the weight function w with the property that for each $n \in S$, $w(n)$ is the probability that 3 occurs first on the nth toss.
    (b)   What is the probability that the die must be tossed at least four times?
    (c)   What is the probability that the 3 will occur on an even-numbered toss?
    (d)   Suppose the 3 occurs on an even-numbered toss, what is the probability that it was the second toss?

2. Prove Proposition 2.

3. Suppose that A and C are independent and B and C are independent. Show that $A \cup B$ and C are independent if and only if $A \cap B$ and C are independent.

4. Give an example of a p.m.s. (S, Σ, P) and three events $\{A_1, A_2, A_3\}$ with the property that

$$P(A_1 \cap A_2 \cap A_3) = P(A_1) \cdot P(A_2) \cdot P(A_3)$$

but $\{A_1, A_2, A_3\}$ is not independent.

5. Let (S, Σ, P) be a p.m.s. and $\{A_1, \ldots, A_n, \ldots\}$ be a countably infinite collection of members of Σ such that each finite subcollection is independent. Prove that

$$P\left(\bigcap_1^\infty A_i\right) = \prod_1^\infty P(A_i).$$

(If a_i is a sequence in R, then $\prod_1^\infty a_i = b$ means $\lim_{n \to \infty} (a_1 \cdot a_2 \cdot \ldots \cdot a_n) = b$.)

6. Let (S, Σ, P) be a p.m.s. and $\{A_1, A_2, \ldots, A_n\}$ be a collection in Σ such that

$$P\left(\bigcap_1^n A_i\right) > 0.$$

Prove that

$$P\left(\bigcap_1^n A_i\right) = P(A_1) \cdot P(A_2 \mid A_1)$$

$$\cdot P(A_3 \mid A_1 \cap A_2) \cdot \ldots \cdot P\left(A_n \mid \bigcap_1^{n-1} A_i\right).$$

7. Suppose (S, Σ, P) is a p.m.s. and A, B, and C are events such that A and B are independent and $P(C) > 0$. Are A and B necessarily independent relative to the conditional probability measure $P((\cdot) \mid C)$?

8. Suppose (S, Σ, P) is a p.m.s. and $A \in \Sigma$. Prove that A and $\emptyset$ are independent and A and S are independent.

4.3   Sequences of Independent Events

In this section we shall be concerned with developing some formulae which will facilitate computing various probabilities.

Proposition 1:   Let $\{A_1, \ldots, A_n\}$ be an independent collection in a p.m.s. (S, Σ, P). Then

$$P\left(\bigcup_1^n A_i\right) = 1 - \prod_1^n P(A_i^c) = 1 - \prod_1^n (1 - P(A_i)).$$

Proof: $P\left(\bigcup_1^n A_i\right) = 1 - P\left(\left[\bigcup_1^n A_i\right]^c\right) = 1 - P\left(\bigcap_1^n [A_i]^c\right)$

$$= 1 - \prod_1^n P(A_i^c) \quad \text{since} \quad \{A^c, \ldots, A_n^c\} \quad \text{also}$$

forms an independent collection. $\square$

Proposition 1 is often used to compute probabilities involving "at least one" of the events $A_1, \ldots, A_n$.

Example: A submarine fires four torpedoes at a ship and each torpedo has a probability of 1/5 of hitting the ship. What is the probability that the ship is hit at least once?

For $i = 1, 2, 3$, and 4, let A_i be the event that the ith torpedo hits the ship. We shall assume that these events are independent. Then the probability that at least one hit occurs is

$$P\left(\bigcup_1^4 A_i\right) = 1 - \prod_1^4 \left(1 - \frac{1}{5}\right) = 1 - \prod_1^4 \left(\frac{4}{5}\right) = \frac{369}{625}.$$

Example: In a certain underdeveloped country, the probability of an infant reaching maturity is only 1/3. How many children should a couple have to ensure that the probability of at least one surviving be at least 9/10?

Let A_i be the event that the ith infant survives; we shall assume (probably wrongly) that the events $\{A_1, \ldots, A_n\}$ form an independent collection. Then we seek n such that

$$P\left(\bigcup_1^n A_i\right) \geq \frac{9}{10}.$$

Now

$$P\left(\bigcup_1^n A_i\right) = 1 - \prod_1^n \left(\frac{2}{3}\right).$$

Thus

$$P\left(\bigcup_1^n A_i\right) \geq \frac{9}{10} \Leftrightarrow \left(\frac{2}{3}\right)^n \leq \frac{1}{10}.$$

The smallest such n is $n = 6$.

The following is an immediate corollary of the preceding proposition.

Proposition 2:   *If $\{A_1, \ldots, A_n\}$ is an independent collection in the p.m.s. (S, Σ, P) then the probability that none of these events occur is $\prod_{1}^{n} (1 - P(A_i))$.*

One important situation in which we encounter independent sequences of events is that of *repeated trials*.

THEOREM 1:   Suppose an "experiment" or "trial" has r possible outcomes and $P(\text{outcome of type } i) = p_i$. If the experiment is repeated n times independently then the probability of

$$\text{exactly } k_1 \text{ outcomes of type 1}$$
$$\text{exactly } k_2 \text{ outcomes of type 2}$$
$$\vdots$$
$$\text{exactly } k_r \text{ outcomes of type } r$$

is $\dfrac{n!}{(k_1! \, k_2! \cdots k_r!) p_1^{k_1} p_2^{k_2} \cdots p_r^{k_r}}$; where $k_1 + k_2 + \cdots + k_r = n$.

Proof: For each i, let s_i be the outcome of type i, then there are $n!/(k_1! \cdots k_r!)$ arrangements of n objects of which k_1 are alike (namely s_1's), k_2 are alike (namely s_2's), etc. Each such arrangement has probability $p_1^{k_1} \cdots p_r^{k_r}$. $\square$

In case the trial has only two outcomes the trials above are called *independent Bernoulli trials*; and the formula for the probability of exactly k_1 outcomes of type 1 becomes $\dfrac{n!}{(k_1! \, (n - k_1)!) p_1^{k_1} (1 - p_1)^{n-k_1}}$.

Example: A fair die is tossed six times. What is the probability of exactly three 1's and two 5's?

Here we are interested in three types of outcomes: $S_1 = \{1\}$, $S_2 = \{5\}$, and $S_3 = \{2, 3, 4, 6\}$. We use the preceding theorem and compute $6!/(3! \, 2! \, 1!)(1/6)^3(1/6)^2(4/6)^1$.

Often probabilities concerning repeated trials will involve factorials of large numbers. We shall present two ways of dealing with such computational problems. The first method attacks the factorials themselves. This may be done by use of a table which gives the logarithms of factorials or by use of Stirling's approximation given in the following proposition, the proof of which may be found in A. E. Taylor's *Advanced Calculus*, published by Ginn and Co., 1955, p. 686.

Proposition:

$$\lim_{n \to \infty} \frac{n!}{n^{n+1/2}e^{-n}\sqrt{2\pi}} = 1.$$

Thus, for large n, $n!$ may be approximated by $n^{n+1/2}e^{-n}\sqrt{2\pi}$.

The other method for easing computational strain is to attack the whole term itself. We do this first of all for a special case.

THEOREM 2:   For each $\lambda > 0$, each $k = 0, 1, 2, \ldots$,

$$\lim_{n \to \infty} \frac{n!}{k!\,(n-k)!} \left(\frac{\lambda}{n}\right)^k \left(1 - \frac{\lambda}{n}\right)^{n-k} = e^{-\lambda}\frac{\lambda^k}{k!}.$$

Proof: We first recall that $\lim_{n \to \infty} (1 - \lambda/n)^n = e^{-\lambda}$.

$$\frac{n!}{k!\,(n-k)!} \left(\frac{\lambda}{n}\right)^k \left(1 - \frac{\lambda}{n}\right)^{n-k}$$

$$= \frac{\lambda^k}{k!} \left(1 - \frac{\lambda}{n}\right)^n \left(1 - \frac{\lambda}{n}\right)^{-k} \frac{n(n-1)\cdots(n-k+1)}{n^k};$$

now as $n \to \infty$, $(1 - \lambda/n)^{-k} \to 1$ and $(n(n-1)\cdots(n-k+1))/n^k \to 1$, but $(1 - \lambda/n)^n \to e^{-\lambda}$; thus,

$$\lim_{n \to \infty} \frac{n!}{k!\,(n-k)!} \left(\frac{\lambda}{n}\right)^k \left(1 - \frac{\lambda}{n}\right)^{n-k} = \frac{\lambda^k}{k!} e^{-\lambda}. \quad \square$$

In the case of n independent Bernoulli trials, the probability of exactly k outcomes of type 1 (where the probability of such an outcome on a given trial is p) is $n!/(k!\,(n-k)!)p^k(1-p)^{n-k}$. If we let $\lambda/n = p$, we see that this probability can be approximated by $e^{-np}(np)^k/k!$ for large n.

For more general independent repeated trials, we use the approximation given by the following theorem whose proof is similar to but messier than that of Theorem 2.

THEOREM 3: Let $r \in \mathbb{N}$ and $k_1, k_2, \ldots, k_{r-1}$ be non-negative integers; let $\lambda_1, \lambda_2, \ldots, \lambda_{r-1}$ be positive numbers. Then

$$\lim_{n \to \infty} \frac{n!}{k_1! \cdots k_{r-1}! \left(n - \sum_1^{r-1} k_i\right)!} \left(\frac{\lambda_1}{n}\right)^{k_1} \left(\frac{\lambda_2}{n}\right)^{k_2} \cdots$$
$$\left(1 - \sum_1^{r-1} \frac{\lambda_i}{n}\right)\left(n - \sum_1^{r-1} k_i\right) = \prod_{i=1}^{r-1} \frac{e^{-\lambda} \lambda^{k_i}}{k_i!}.$$

Now the formula in Theorem 1, namely $n!/(k_1! \cdots k_r!)p_1^{k_1} \cdots p_r^{k_r}$, may be approximated by

$$\prod_{i=1}^{r-1} e^{-np_i} \frac{(np_i)^{k_i}}{(k_i)!}$$

for large n.

Example: A machine produces light bulbs of which 10 percent are defective. If a sample of 100 is chosen independently what is the probability that exactly 10 are defective?

The answer by Theorem 1 is $A = 100!/(10!\ 90!)(1/10)^{10}(9/10)^{90}$. We shall approximate this by each of our methods.

Method I: Use of Logarithms.
We find that

$\log_{10}(100!) = 157.97000$	$\log_{10}(1/10)^{10} = 10.00000$
$\log_{10}(90!) = 138.17194$	$\log_{10}(9/10)^{90} = -4.11840$
$\log_{10}(10!) = 6.55976$	

Thus $\log_{10}(A) = -0.88010$ so $A = 0.1318$.

Method II: Stirling's Approximations of Factorials

$$A \doteq \frac{100^{100+1/2} \sqrt{2\pi}\ e^{-100}}{10^{10+1/2} \sqrt{2\pi}\ e^{-10}\ 90^{90+1/2} \sqrt{2\pi}\ e^{-90}} \left(\frac{1}{10}\right)^{10} \left(\frac{9}{10}\right)^{90}$$

$$= \frac{(100)^{100} \cdot 10}{10^{10} \cdot \sqrt{10} \sqrt{2\pi}\ 90^{90} \sqrt{90}\ 10^{100}} \frac{9^{90}}{}$$

$$= \frac{(100)^{100} \cdot 10}{10^{110} \cdot 10 \cdot 3 \sqrt{2\pi}\ 9^{90} \cdot 10^{90}} 9^{90}$$

$$= \frac{1}{3\sqrt{2\pi}} \left(\frac{(100)^{100}}{(10)^{200}}\right) = \frac{1}{3\sqrt{2\pi}} \doteq \frac{1}{7.520} \doteq 0.1329$$

Method III: Exponential Approximation

$$A \doteq \frac{e^{-10}10^{10}}{10!} = \frac{e^{-10}10^9}{9!} = \frac{4.54 \times 10^{-5} \times 10^9}{3.63 \times 10^5}$$

$$\doteq 1.25 \times 10^{-1} = 0.125.$$

The approximation used in Method III above is interesting in its own right. The probability measure on $\mathscr{P}(\{0, 1, \ldots, n, \ldots\})$ which is generated by the weight function

$$w(n) = \frac{e^{-\lambda}\lambda^n}{n!} \quad (\lambda > 0)$$

is called the *Poisson measure with parameter* λ. It has been discovered relatively recently that many random phenomena may be modeled by this probability measure.

Example: A particular type of insect lays eggs at most once a year. For $k = 0, 1, 2, \ldots$, the probability that a female insect lays k eggs is $e^{-\lambda}\lambda^k/k!$. Each egg has a probability of p of hatching and the insect reaching maturity. Investigate the number of mature insects produced by a given female.

We shall assume that survival of the eggs are independent.

$$P(m \text{ mature}) = P(m \text{ mature} \cap \geq m \text{ produced})$$

$$= \sum_{k=m}^{\infty} P(m \text{ mature} \cap k \text{ produced})$$

$$= \sum_{k=m}^{\infty} P(m \text{ mature} \mid k \text{ produced}) \cdot P(k \text{ produced})$$

$$= \sum_{k=m}^{\infty} \frac{k!}{m! \, (k-m)!} \, p^m (1-p)^{k-m} \cdot \frac{e^{-\lambda}\lambda^k}{k!}$$

$$= \frac{p^m e^{-\lambda}\lambda^m}{m!} \sum_{j=0}^{\infty} \frac{[\lambda(1-p)]^j}{j!} \quad \text{where } j = k - m$$

$$= \frac{(\lambda p)^m e^{-\lambda}}{m!} \cdot e^{\lambda(1-p)}$$

$$= \frac{e^{-\lambda p}(\lambda p)^m}{m!}.$$

Exercises:

1. Four riflemen shoot independently at a target. The first rifleman has a probability of 1/2 of hitting it, the second, 3/4, and the other two each have a probability of 1/4. What is the probability that
   (a) The target will be hit at least once?
   (b) The target will be hit exactly once?

2. In a well-established method of operation, a machine produces fuses of which 1/20 are defective. An employee claims that a new method will increase production and not raise the number of defectives. However, when the new method is employed for a substantial period of time and a sample of ten independent fuses is tested, it is found that one is defective. Should the new method be dropped?

3. A nuclear engineer is faced with ordering safety devices for a nuclear reactor. There are two types available; type A fails 10 percent of the time and type B fails 1 percent of the time. Due to the possibility of a melt-down, it is required that the probability of all failing be at most 10^{-10}. Assume that the operations of the devices are independent.
   (a) If only devices of type A are to be installed, how many should he order?
   (b) If only devices of type B, how many?
   (c) In a test of one device of type A and one of type B, it was observed that one failed and the other worked. What is the probability that the one that failed was of type A?

4. A machine produces large quantities of a particular item but 10 percent are defective. A sample of ten is picked at random from the output of the machine. What is the probability that none are defective?

5. Fred has one coin and John has two. They will match coins until one has all three coins.
   (a) For each $n \in N$, find the probability that the game ends on the nth toss.
   (b) Show that Fred wins all three coins if and only if the game ends on an even toss.
   (c) Find the probability that Fred wins all three coins.

5

Random Variables

5.1 Introduction and Definition

Often in investigating random phenomena there are numbers involved in the events whose probabilities we seek. For example, in a toss of five coins, we may be interested in the probability of getting three heads. We might take for a sample space S, all ordered 5-tuples $(a_1, \ldots, a_5)$ where each a_i is either "H" or "T." Then to each member of the sample space, we associate a number, namely the number of heads in that member. This defines a function X from the sample space S into the real numbers (into $\{0, 1, 2, 3, 4, 5\}$, in fact). The probability of three heads is then $P\{s \in S \mid X(s) = 3\}$. There can be exactly three heads in $5!/3!\,2!$ ways. If we assume that the coins are fair and the outcomes are independent, then each member $(a_1, \ldots, a_5)$ of S has probability $(1/2) \cdot (1/2) \cdot (1/2) \cdot (1/2) \cdot (1/2) = 1/2^5$; so $P\{s \in S \mid X(s) = 3\} = (5!/3!\,2!)\,(1/2)^5$. Of course, this problem was worked using the methods of the preceding chapter—our only reason for using it is to introduce the concept of a real valued function on a sample space.

We shall be concerned not necessarily with all real valued functions on a sample space, but with those for which we can compute certain important probabilities.

Definition 1:   Let (S, Σ, P) be a p.m.s. and X a function from S into the real numbers $(X: S \to \mathsf{R})$. X is called a *random variable* (usually abbreviated *r.v.*) provided that for each $a \in \mathsf{R}$,

$$\{s \in S \mid X(s) \leq a\} \in \Sigma.$$

The important fact about a random variable is that for each real number a, we may compute the probability that X be less than or equal to a.

Example: Let S be any set and $\Sigma = \mathscr{P}(S)$; let P be arbitrary.

Any function $X: S \to \mathsf{R}$ is a random variable since $\{s \in S \mid X(s) \leq a\}$ is always a subset of S and each subset of S is a member of Σ.

Example: Let $S = \mathsf{R}$ and Σ be the collection of all sets $A \subseteq S$ such that either A is countable or $S \backslash A$ is countable; let P be arbitrary.

The function $X: S \to \mathsf{R}$ by $X(s) = s$ is not a r.v. since

$$\{s \in S \mid X(s) \leq 0\} = (-\infty, 0];$$

neither $(-\infty, 0]$ nor its complement is countable so $(-\infty, 0] \notin \Sigma$. The function $Y: S \to \mathsf{R}$ by $Y(s) = \begin{cases} 1 & \text{if } s \in \mathsf{Q} \\ 0 & \text{if } s \notin \mathsf{Q} \end{cases}$ is a r.v. since

for $a \geq 1$, $\{s \in S \mid Y(s) \leq a\} = S \in \Sigma$

for $0 \leq a < 1$, $\{s \in S \mid Y(s) \leq a\} = S \backslash \mathsf{Q} \in \Sigma$

for $a < 0$, $\{s \in S \mid Y(s) \leq a\} = \varnothing \in \Sigma$.

Example: A coin will be tossed until a head is obtained. You will be awarded \$2.00 if the head occurs on an even-numbered toss and nothing if the head occurs on an odd-numbered toss.

We first set up a p.m.s. that models the coin-tossing process. Let $S = \mathsf{N}$ and $\Sigma = \mathscr{P}(S)$; we define a weight function w on S such that

$w(n)$ is the probability that the head occurs on the nth toss. Since this happens only if the first $n - 1$ tosses result in tails and the nth toss in heads, then

$$w(n) = (1/2)^{n-1} \cdot (1/2) = 1/2^n.$$

Now let X be the amount you receive, for example, $X(1) = 0$, $X(2) = 2$, $X(3) = 0$, $X(4) = 2, \ldots$. In general,

$$X(n) = \begin{cases} 2 & \text{if } n \text{ is even} \\ 0 & \text{if } n \text{ is odd} \end{cases}$$

One probability of interest is the probability of payoff:

$$P(\{s \in S \mid X(s) = 2\}) = P(\{2, 4, 6, 8, \ldots\})$$

$$= \sum_{n=1}^{\infty} w(2n) = \sum_{n=1}^{\infty} 1/2^{2n}$$

$$= \sum_{n=1}^{\infty} 1/4^n = \frac{1/4}{1 - 1/4} = 1/3.$$

For a r.v. X we may compute $P\{s: X(s) \le a\}$ for each a, but there may be other questions to which we want answers; for example, we may want to know the probability that X be equal to π ($P\{s \in S \mid X(s) = \pi\}$) or the probability that X be a rational number ($P\{s \in S \mid X(s) \in Q\}$). We shall show that each of these probabilities and many more can be computed for a r.v.

Definition 2: Let A and B be sets and $f: A \to B$. Then

   (1)   for each $b \in B$, $f^{-1}(b) = \{a \in A \mid f(a) = b\}$,

   (2)   for each $C \subseteq B$, $f^{-1}(C) = \{a \in A \mid f(a) \in C\}$.

$f^{-1}(C)$ is called the *inverse image* (pre-image) of C.

We stress, that in general, f^{-1} is not a function from B into A (although it may be considered a function from B into $\mathscr{P}(A)$).

Example: Let $A = R = B$ and $f(x) = x^2$.

Then $f^{-1}(0) = \{0\}$, $f^{-1}(1) = \{+1, -1\}$, $f^{-1}(-1) = \varnothing$, $f^{-1}((-4, 4]) = [-2, 2]$, $f^{-1}([-728, 4]) = [-2, 2]$.

The condition "$\{s \in S \mid X(s) \leq a\} \in \Sigma$" in the definition of a r.v. can be written "$\{s \in S \mid X(s) \in (-\infty, a]\} \in \Sigma$" which can be written in the notation of the preceding definition as "$X^{-1}((-\infty, a]) \in \Sigma$." The next two theorems indicate some properties of those subsets B of R with the property that $X^{-1}(B) \in \Sigma$.

THEOREM 1:   Let Σ be a σ-algebra of subsets of a set S and let $X: S \to R$ be any function. Then the class of all subsets B of R with the property that $X^{-1}(B) \in \Sigma$ forms a σ-algebra of subsets of R.

Proof: Let $\mathcal{M}$ be the collection of all $B \subseteq R$ such that $X^{-1}(B) \in \Sigma$. We shall show that $\mathcal{M}$ is closed under countable union and under complement and thus is a σ-algebra. If $\{A_i\}_1^\infty$ is a sequence in $\mathcal{M}$ then $X^{-1}(A_i) \in \Sigma$ for each i. Let $A = \bigcup_1^\infty A_i$, then

$$X^{-1}(A) = \left\{ s \in S \mid X(s) \in A = \bigcup_1^\infty A_i \right\} = \{s \in S \mid X(s) \in A_i \text{ for some } i\}$$

$$= \bigcup_1^\infty \{s \in S \mid X(s) \in A_i\} = \bigcup_1^\infty X^{-1}(A_i).$$

The last set above is a member of Σ since Σ is a σ-algebra. If $A \in \mathcal{M}$, then

$$X^{-1}(A^c) = \{s \in S \mid X(s) \in A^c\} = \{s \in S \mid X(s) \notin A\} = \{s \in S \mid X(s) \in A\}^c$$

$$= [X^{-1}(A)]^c \in \Sigma. \quad \square$$

We recall that the Borel sets in R form the smallest σ-algebra $\mathcal{B}$ containing all sets of the form $(-\infty, a]$.

THEOREM 2:   Let (S, Σ, P) be a p.m.s. and $X: S \to R$ be a r.v. For each Borel set $B \subseteq R$, $X^{-1}(B) \in \Sigma$.

Proof: By the definition of r.v., $X^{-1}((-\infty, a]) \in \Sigma$ for each $a \in R$. Now if $\mathcal{M}$ is the collection of all subsets B of R such that $X^{-1}(B) \in \Sigma$, then $\mathcal{M}$ is a σ-algebra (by Theorem 1) which contains all sets of the form $(-\infty, a]$. Thus $\mathcal{M}$ contains the *smallest* σ-algebra which contains all such sets. This smallest σ-algebra is $\mathcal{B}(R)$. $\quad \square$

Theorem 2 assures us that if X is a r.v., then $P(\{s \in S \mid X(s) \in B\})$ may be computed whenever B is a Borel set. Of course, that *actual* computation may be extremely difficult but we do know that $\{s \in S \mid X(s) \in B\} \in \Sigma$ so it has a probability. The next section will include a method of computation for some simple Borel sets.

We have the following immediate corollary to Theorem 2.

Corollary 1: Let (S, Σ, P) be a p.m.s. and let $X: S \to \mathrm{R}$. These are equivalent:

    (i) X is a r.v.

    (ii)   For each Borel set $B \subseteq \mathrm{R}$, $X^{-1}(B) \in \Sigma$

    (iii)   For each set of the form (a, b), $X^{-1}((a, b)) \in \Sigma$.

The reader is invited to concoct similar equivalent statements based on the many ways that $\mathscr{B}(\mathrm{R})$ may be generated. (See Theorem 2 of chapter 2, section 3.)

We shall see later that a random variable may be composed with a very general function from R to R and the result is still a random variable. We mention two special cases below.

Proposition 1: *If X is a r.v. on the p.m.s. (S, Σ, P), then*

    (i)   so is $Y(s) = [X(s)]^2$

    (ii)   so is $Z(s) = aX(s) + b$ where $a, b \in \mathrm{R}$.

Proof:

    (i)   We shall show that if $t \in \mathrm{R}$, then $\{s \in S \mid Y(s) \le t\} \in \Sigma$. If $t < 0$, then

$$\{s \in S \mid Y(s) \le t\} = \{s \in S \mid [X(s)]^2 \le t < 0\} = \emptyset \in \Sigma.$$

If $t \ge 0$, then

$$
\begin{aligned}
\{s \in S \mid Y(s) \le t\} &= \{s \in S \mid [X(s)]^2 \le t\} \\
&= \{s \in S \mid -\sqrt{t} \le X(s) \le \sqrt{t}\} \\
&= X^{-1}([-\sqrt{t}, \sqrt{t}])
\end{aligned}
$$

which is in Σ since $[-\sqrt{t}, \sqrt{t}]$ is a Borel set.

(ii)   We shall prove (ii) only in the case in which $a < 0$, leaving the
other cases to the reader.

$$\{s \in S \mid Z(s) \leq t\} = \{s \in S \mid aX(s) + b \leq t\}$$
$$= \{s \in S \mid aX(s) \leq t - b\}$$
$$= \{s \in S \mid X(s) \geq (t - b)/a\}$$
$$= X^{-1}([(t - b)/a, \infty))$$

which is in Σ since $[(t - b)/a, \infty)$ is a Borel set.

Exercises:

*1.   Let (S, Σ, P) be a p.m.s. and $A \subseteq S$. Let

$$\chi_A: S \to R \quad \text{by} \quad \chi_A(s) = \begin{cases} 1 & \text{if } s \in A \\ 0 & \text{if } s \notin A \end{cases}$$

(χ_A is called the *characteristic function* of the set A.) Prove that χ_A is a
random variable if and only if $A \in \Sigma$.

2.   Let (S, Σ, P) be a p.m.s. and $A_1, \ldots, A_n$ members of Σ. Define
$X: S \to R$ by $X(s) =$ number of A_i's which contain s. Is X a r.v.?

3.   Let X be a r.v. on a p.m.s. (S, Σ, P) and $g: R \to R$ such that $t_1 \leq t_2$
$\Rightarrow g(t_1) \leq g(t_2)$. Prove that $Y: S \to R$ by $Y(s) = g \circ X(s) = g[X(s)]$ is a r.v.

4.   Let A and B be sets and $f: A \to B$. Prove

(i)   if $\{B_\alpha\}_{\alpha \in \mathfrak{A}}$ is any collection of subsets of B then

(a) $f^{-1}\left(\bigcup_{\mathfrak{A}} B_\alpha\right) = \bigcup_{\mathfrak{A}} f^{-1}(B_\alpha)$

(b) $f^{-1}\left(\bigcap_{\mathfrak{A}} B_\alpha\right) = \bigcap_{\mathfrak{A}} f^{-1}(B_\alpha)$

(ii)   if $B_1, B_2 \subseteq B$, then $f^{-1}(B_1 \backslash B_2) = f^{-1}(B_1) \backslash f^{-1}(B_2)$.

5.   Let X be a r.v. on a p.m.s. (S, Σ, P), indicate why $\{s \in S \mid X(s)$ is
rational$\}$ is an event.

6.   Accept the invitation extended immediately after Corollary 1.

7. Prove part (ii) of Proposition 1 in the remaining cases.

8. X is a r.v. on a p.m.s. (S, Σ, P). Let $Y: S \to \mathsf{R}$ by $Y(s) = e^{X(s)}$. Prove that Y is a r.v.

9. A fair coin is tossed either five times or until a head occurs (whichever is first). X is the number of times the coin is tossed. Set up a p.m.s. on which X can be defined as a r.v. Find $P(\{s \in S \mid X(s) = 5\})$.

5.2   Distribution Functions

In the preceding section we were concerned with a random variable X on a p.m.s. (S, Σ, P) but practically nothing was mentioned about the probability measure P. In this section, we tie together the r.v. and P.

Definition 1:   Let X be a r.v. on a p.m.s. (S, Σ, P). Define $F_X: \mathsf{R} \to \mathsf{R}$ by $F_X(t) = P\{s \in S \mid X(s) \leq t\}$; then F_X is called the (cumulative) *distribution function* of X.

First, note that if X is a r.v., then for each $t \in \mathsf{R}$, $\{s \in S \mid X(s) \leq t\} \in \Sigma$; thus $P\{s \in S \mid X(s) \leq t\}$ is defined. Second, note that while X is a function from S into R, F_X is a function from R into R. The random variable X itself may be difficult to analyze (how would you draw the graph, for example?), however, we shall find that the distribution function will be much easier to analyze and will tell us a great deal about X. We shall abbreviate "$P\{s \in S \mid X(s) \leq t\}$" by "$P(X \leq t)$" which is usually read "the probability that X be less than or equal to t." Similarly we shall write $P(X \geq t)$ and $P(X = t)$; for a Borel set B we shall write $P(X \in B)$.

Example: Flip two fair coins independently and let X be the number of heads obtained.

Here we take $S = \{(H, H), (H, T), (T, H), (T, T)\}$, $\Sigma = \mathscr{P}(S)$, and P, the equiprobable measure. We have that $X(H, H) = 2, X(H, T) = X(T, H) = 1$, and $X(T, T) = 0$. The distribution function for X is

$$F_X(t) = P(X \leq t) = \begin{cases} P(\varnothing) & \text{if } t < 0 \\ P((T, T)) & \text{if } 0 \leq t < 1 \\ P((T, T), (H, T), (T, H)) & \text{if } 1 \leq t < 2 \\ P(S) & \text{if } t \geq 2 \end{cases}$$

$$\begin{cases} 0 & \text{if } t < 0 \\ 1/4 & \text{if } 0 \leq t < 1 \\ 3/4 & \text{if } 1 \leq t < 2 \\ 1 & \text{if } t \geq 2 \end{cases}$$

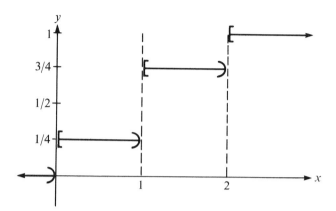

Figure 5.1:   Graph of F_X

Before looking at various types of distribution functions, we list some properties which all distribution functions possess.

THEOREM 1:   Let F_X be the distribution function of a r.v. X on a p.m.s. (S, Σ, P). Then

(a) $0 \le F_X(x) \le 1$ for each $x \in \mathsf{R}$

(b)   if $x \le y$, then $F_X(x) \le F_X(y)$

(c) $\lim_{x \to \infty} F_X(x) = 1$

(d) $\lim_{x \to -\infty} F_X(x) = 0$

(e) $\lim_{x \to a^+} F_X(x) = F_X(a)$ for each $a \in \mathsf{R}$.

Proof:

(a)   Since $F_X(x) = P(X \le x)$ and for each $A \in \Sigma$, $0 \le P(A) \le 1$, then $0 \le F_X(x) \le 1$.

(b)   If $x \le y$, then $\{s \in S \mid X(s) \le x\} \subseteq \{s \in S \mid X(s) \le y\}$. Thus $F_X(x) = P(X \le x) \le P(X \le y) = F_X(y)$.

(c)   For each $n \in \mathsf{N}$, let $A_n = \{s \in S \mid X(s) \le n\}$; then $P(A_n) = F_X(n)$. Now for each $s \in S$, $X(s) \in \mathsf{R}$ so there is an integer n such that $X(s) \le n$; thus $s \in A_n$ and so $S = \bigcup_{1}^{\infty} A_n$.   Since $A_1 \subseteq$

$A_2 \subseteq \cdots$, we may use one of the continuity properties of the probability measure P to deduce that

$$1 = P(S) = P\left(\bigcup_1^\infty A_n\right) = \lim_{n\to\infty} P(A_n) = \lim_{n\to\infty} F_X(n).$$

This result along with (b) ensures (c).

(d) The proof of (d) is similar to that of (c) except that we let $A_n = \{s \in S \mid X(s) \leq -n\}$ and use the other continuity property of P.

(e) Let $a \in R$ and for each $n \in N$, let $A_n = X^{-1}((-\infty, a + 1/n])$; then $A_1 \supseteq A_2 \supseteq \cdots$, $\bigcap_1^\infty A_n = X^{-1}((-\infty, a])$ and $P(A_n) = F_X(a + 1/n)$. Thus $F_X(a) = P(X^{-1}(-\infty, a]) = \lim_{n\to\infty} P(A_n) = \lim_{n\to\infty} F_X(a + 1/n)$ and (e) follows from (b). $\square$

The reader may have noticed a similarity between the concept of a distribution function for a random variable and a distribution function for a probability measure on the Borel subsets of R. In fact, the two concepts coincide exactly.

THEOREM 2: Let $F: R \to R$; then these are equivalent:

(a) F is the distribution function for some probability measure on the Borel sets $\mathscr{B}(R)$ of R.

(b) F is the distribution function for some random variable X on some p.m.s. (S, Σ, P).

(c) F has the properties
   (i) $0 \leq F(x) \leq 1$ for each $x \in R$
   (ii) if $x \leq y$, then $F(x) \leq F(y)$
   (iii) $\lim_{x\to\infty} F(x) = 1$
   (iv) $\lim_{x\to-\infty} F(x) = 0$
   (v) $\lim_{x\to a^+} F(x) = F(a)$ for each $a \in R$.

Proof: (*a* ⇒ *b*)   Let $S = \mathsf{R}$, $\Sigma = \mathscr{B}(\mathsf{R})$, and P be as given in (a). Define $X: S \to \mathsf{R}$ by $X(s) = s$. Then $F_X(t) = P(X \le t) = P((-\infty, t]) = F(t)$.

(*b* ⇒ *c*)   This was done by Theorem 1.

(*c* ⇒ *a*)   This was taken care of by Theorem 3 of chapter 2, section 3.   □

Perhaps the most interesting aspect of Theorem 2 is that each abstract random variable X generates a random variable Y on $(\mathsf{R}, \mathscr{B}(\mathsf{R}))$ such that $F_X(t) = F_Y(t)$ for each $t \in \mathsf{R}$. Actually, the random variable generated is trivial: $Y(s) = s$; it is the probability measure on $\mathscr{B}(\mathsf{R})$ which is the key. Theorem 2 ensures that if $B \in \mathscr{B}(\mathsf{R})$, then $X^{-1}(B)$ is an event. We define $P^*(B) = P(X^{-1}(B))$. We leave as an exercise to prove directly that P^* is a probability measure on $\mathscr{B}(\mathsf{R})$. We shall call any function satisfying properties (i)–(v) of the preceding theorem, a *distribution function*.

We note from Theorems 1 and 2 that a distribution function F has the property that $\lim_{x \to a^+} F(x) = F(a)$; that is, F is continuous from the right at each real number. However, F may not necessarily be continuous everywhere on R. For example, if (S, Σ, P) is a p.m.s. and $A \in \Sigma$, then $X = \chi_A$ is a r.v. (see exercise 1 of the preceding section). The distribution function F_X has the graph shown in figure 5.2.

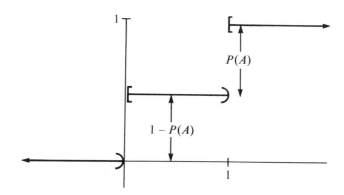

Figure 5.2:   Distribution Function of χ_A

Since each distribution function F is non-decreasing $\lim_{x \to a^-} F(x)$ always exists; we shall denote this limit by $F(a^-)$.

The following theorem is very useful in computing probabilities involving random variables. The proof will be left as an exercise (see Theorem 4, chapter 2, section 3).

THEOREM 3: Let F_X be the distribution function of a random variable X on a p.m.s. (S, Σ, P); let $a, b \in \mathsf{R}$ with $a \le b$. Then

(i) $P(X > a) = 1 - F_X(a)$

(ii) $P(X < a) = F_X(a^-)$

(iii) $P(a < X \le b) = F_X(b) - F_X(a)$

(iv) $P(a \le X \le b) = F_X(b) - F_X(a^-)$

(v) $P(X = a) = F_X(a) - F_X(a^-)$

(vi) $P(a < X < b) = F_X(b^-) - F_X(a)$ if $a < b$

(vii) $P(a \le X < b) = F_X(b^-) - F_X(a^-)$ if $a < b$.

Example: The interval $[0, 2]$ will be cut at random in the sense that the probability of the cut point falling in a particular subinterval of $[0, 2]$ is proportional to the length of the subinterval.

For example, $P(X \in (1/2, 1)) = P(X \in [0, 1]) \cdot 1/2$. Now

$$F_X(t) = P(X \le t) = P(X \in [0, t])$$

$$= \frac{\text{length } [0, t]}{\text{length } [0, 2]} = t/2 \quad \text{if} \quad 0 \le t \le 2;$$

with $F_X(t) = 0$ if $t < 0$ and $F_X(t) = 1$ if $t > 2$. Hence

$$F_X(t) = \begin{cases} 0 & \text{if } t < 0 \\ t/2 & \text{if } 0 \le t < 2 \\ 1 & \text{if } t \ge 2 \end{cases}$$

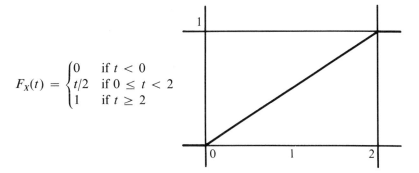

Now we find the probability that the cut point be less than 1, for example, to be $F_X(1^-) = F_X(1) = 1/2$. Let us compute the probability that the longer subinterval formed by the cut be less than or equal to t. Let Y be the length of the longer subinterval. One subinterval has length X and the other, $2 - X$. Thus $P(Y \le t) = P$(both subintervals have length $\leq t$) $= P(X \leq t$ and $2 - X \leq t)$. Now, certainly if $t < 1$, then $P(Y \leq t) = 0$ and if $t > 2$, then $P(Y \leq t) = 1$. If $1 \leq t \leq 2$, then $P(Y \leq t) = P(X \leq t$ and $2 - X \leq t) = P(2 - t \leq X \leq t) = F_X(t) - F_X((2 - t)^-) = t/2 - (2 - t)/2 = t - 1$ and the graph of F_Y is given in figure 5.3.

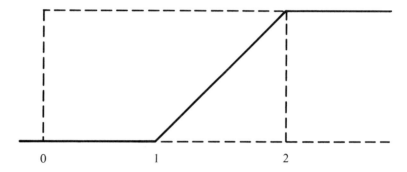

Figure 5.3

Exercises:

1. Let X be a r.v. and $a, b \in R$ with $a > 0$. Let $Y(s) = a \cdot X(s) + b$. Find F_Y in terms of F_X, a, and b.

2. Let X be a r.v. and $Y(s) = [X(s)]^2$. Find F_Y in terms of F_X.

3. Let X be a r.v. on the p.m.s. (S, Σ, P). For each Borel set $B \subseteq R$, let $P^*(B) = P(X^{-1}(B))$. Prove that P^* is a probability measure on $\mathcal{B}(R)$.

4. Prove Theorem 3.

5. For each of the functions given below, (1) sketch the function, (2) convince yourself that it is a distribution function for a r.v. X, (3) find $P(-1 \leq X \leq 1)$, and (4) find $P(-1 \leq X \mid X \leq 1)$.

   (a) $F(x) = \begin{cases} 0 & \text{if } x < 0 \\ x^2/2 & \text{if } 0 \leq x < 1 \\ 1 & \text{if } x \geq 1 \end{cases}$

   (b) Let $[x]$ denote the largest integer n which is less than or equal to x.
   $$F(x) = \begin{cases} 0 & \text{if } x < 0 \\ 1 - (1/2)^{[x]} & \text{if } x \geq 0 \end{cases}$$

(c) $F(x) = \begin{cases} (1/2)e^x & \text{if } x < 0 \\ 1 - (1/4)e^{-x} & \text{if } x \geq 0 \end{cases}$

(d) $F(x) = 1/\pi(\text{Tan}^{-1} x + \pi/2)$ (recall $-(\pi/2) < \text{Tan}^{-1}(x) < \pi/2$)

*6. Suppose F_1 and F_2 are distribution functions and $0 \leq \lambda \leq 1$. Prove that $\lambda F_1 + (1 - \lambda)F_2$ is a distribution function. (Hint: use Theorem 2.)

5.3 Classification of Random Variables

If F is a distribution function, then we know that F is continuous from the right on R and we have seen examples in which F is not continuous everywhere on R. F is non-decreasing so, as the following result shows, it can be discontinuous at, at most, a countable number of points.

Proposition 1:   *Let $F: R \to R$ such that*

(a) *F is non-decreasing*

(b) *F is bounded above (by, say M)*

(c) *F is bounded below (by, say m).*

Then the set of points of discontinuity of F is countable.

Proof: We shall first show that $F(a^+) = \lim\limits_{x \to a^+} F(x)$ exists for each $a \in R$; a similar proof shows that $F(a^-)$ exists. The set $\{F(x) \mid x > a\}$ is bounded below by $F(a)$ and thus has a greatest lower bound $b \geq F(a)$. Since F is non-decreasing we may pick a sequence $\{x_n\}_1^\infty$ such that $x_n \to a$, $x_1 > x_2 > x_3 \cdots$, and $\lim\limits_{n \to \infty} F(x_n) = b$. Now if $\varepsilon > 0$, then there is a positive integer N such that if $n \geq N$, then $|F(x_n) - b| < \varepsilon$. Let $\delta = x_N - a > 0$; if $a < x < a + \delta$, then $a < x < x_N$ so $b < F(x) < F(x_N)$. Thus $|F(x) - b| < \varepsilon$ and so $\lim\limits_{x \to a^+} F(x) = b$.

For each $x \in R$, let $J(x) = F(x^+) - F(x^-)$, then $J(x)$ is the jump of F at x. Let

$$A_1 = \{x \in R \mid J(x) \geq 1\},$$

then A_1 must be finite; indeed, there can be no more than $M - m$ elements of A since the total rise of F is $M - m$. Let

$$A_2 = \{x \in R \mid J(x) \geq 1/2\},$$

then A_2 must be finite (no more than $2 \cdot (M - m)$ elements). For each $n \in N$, the set

$$A_n = \{x \in R \mid J(x) \geq 1/n\}$$

is finite (with no more than $n \cdot (M - m)$ elements.) The function F is discontinuous at x if and only if $J(x) > 0$ which occurs exactly when $x \in A_n$ for some $n \in N$. Thus the set of points of discontinuity of F is exactly $\bigcup_1^\infty A_n$ which is countable. $\square$

A distribution function satisfies the hypotheses of the preceding proposition so it may be discontinuous only on a countable set. The total rise of a distribution function is 1 and we single out the particular case in which the rise is accounted for entirely by jumps.

Definition 1:   Let F_X be the distribution function of a r.v. X on a p.m.s. (S, Σ, P).   For each $t \in R$, let $p_X(t) = F_X(t) - F_X(t^-)$; then $p_X(t)$ is called the jump of F at t. If F_X can be expressed as

$$F_X(x) = \sum_{t \leq x} p_X(t)$$

for each $x \in R$, then X is called a *discrete* random variable and p_X is called the *probability mass function* for X. Sometimes it is said that F_X is discrete.

We note that $p_X(t) = P(X = t)$ for each $t \in R$ and that for a discrete random variable, there is a *countable* collection $\{a_1, a_2, \ldots, a_n, \ldots\}$ of real numbers such that $\sum_{i=1}^\infty p_X(a_i) = 1$. (Caution: there may be only a finite number of a_i's for which we actually have $p_X(a_i) > 0$.) Conversely if X is any random variable for which there exists such a collection, then X is discrete. Thus any function $p: R \rightarrow [0, \infty)$ such that $\{a \in R \mid p(a) > 0\}$ is countable and $\sum_{a \in R} p(a) = 1$ generates a distribution function for a discrete random variable by the formula

$$F(x) = \sum_{t \leq x} p(t).$$

If X is discrete and a and b are real numbers with $a < b$ such that if $a < t < b$, then $p_X(t) = 0$; then F_X is constant on $[a, b)$; conversely, if F_X is constant on $[a, b)$, then $p_X(t) = 0$ for each $t \in (a, b)$.

Example: Investigate the number of heads in four tosses of a fair coin. Let $S_1 = S_2 = S_3 = S_4 = \{H, T\}$ and $S = S_1 \times S_2 \times S_3 \times S_4$; let $\Sigma = \mathscr{P}(S)$ and P be the equiprobable measure on Σ. We are interested in the r.v. $X: S \to \mathsf{R}$ by $X(a_1, a_2, a_3, a_4) = $ number of heads in (a_1, a_2, a_3, a_4). We shall derive the probability mass function p_X and then the distribution function F_X. Now $p_X(t) = P(X = t)$ so $p_X(t) = 0$ unless $t = 0, 1, 2, 3,$ or 4. For these values, we compute

$$p_X(0) = P(X = 0) = 1/2^4$$
$$p_X(1) = P(X = 1) = 4/2^4$$
$$p_X(2) = P(X = 2) = 6/2^4$$
$$p_X(3) = P(X = 3) = 4/2^4$$
$$p_X(4) = P(X = 4) = 1/2^4.$$

The function p_X *is* a probability mass function since $p_X(0) + p_X(1) + p_X(2) + p_X(3) + p_X(4) = 1$. The graph of p_X is given in figure 5.4.

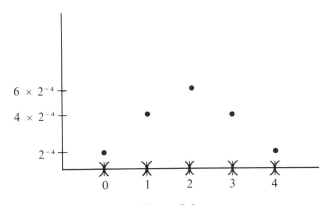

Figure 5.4

The distribution function F_X then may be computed:

$$F_X(t) = \begin{cases} 0 & \text{if } t < 0 \\ 1/2^4 & \text{if } 0 \le t < 1 \\ 1/2^4 + 4/2^4 = 5/2^4 & \text{if } 1 \le t < 2 \\ 5/2^4 + 6/2^4 = 11/2^4 & \text{if } 2 \le t < 3 \\ 11/2^4 + 4/2^4 = 15/2^4 & \text{if } 3 \le t < 4 \\ 15/2^4 + 1/2^4 = 1 & \text{if } t \ge 4 \end{cases}$$

and the graph of F_X is given in figure 5.5.

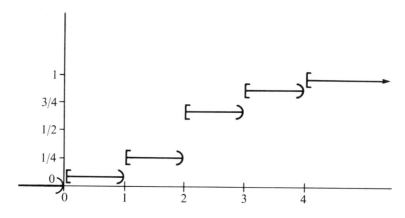

Figure 5.5

Example: A fair coin is tossed until a head is obtained. Investigate the number of tosses required.

Let $S = N$, $\Sigma = \mathscr{P}(N)$ and let $X: S \to R$ by $X(s) = s$. The important part of this p.m.s. is the probability measure which will be generated by a weight function. Since $X(s) = s$, then $p_X(t) = P(X = t) = P(\{t\})$ for $t \in S = N$ and $p_X(t) = 0$ if $t \notin N$. $X = n$ when the head is first obtained on the nth toss; this happens only when the first $n - 1$ tosses result in tails and the nth toss in heads. This has probability $(1/2)^{n-1} \cdot (1/2) = 1/2^n$, so

$$p_X(t) = \begin{cases} 1/2^n & \text{if } t = n \in N \\ 0 & \text{if } t \notin N \end{cases}.$$

The graph of p_X is indicated in figure 5.6.

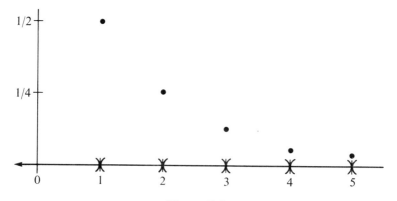

Figure 5.6

The distribution function is given by

$$F_X(t) = \begin{cases} 0 & \text{if } t < 1 \\ 1/2 & \text{if } 1 \le t < 2 \\ 3/4 & \text{if } 2 \le t < 3 \\ \vdots & \\ 1 - 1/2^n & \text{if } n \le t < n + 1 \\ \vdots & \end{cases}$$

and the graph of F_X is indicated in figure 5.7.

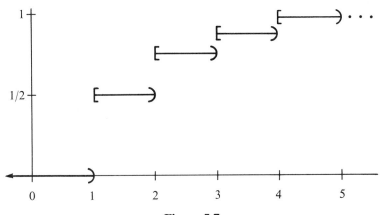

Figure 5.7

Example: A lot of 120 fuses contains 10 defective fuses and a random sample of 6 fuses is chosen.

We shall investigate the number of defective fuses in the sample. Let X be the number of such defectives. Then if t is neither $0, 1, 2, \ldots, 6$ then $P(X = t) = 0$, whereas if $t = 0, 1, 2, \ldots, 6$ then

$$P(X = t) = \frac{C(10, t) \cdot C(110, 6 - t)}{C(120, 6)}.$$

Thus

$$p_X(t) = \begin{cases} \dfrac{C(10, t)C(110, 6 - t)}{C(120, 6)} & \text{if } t = 0, 1, 2, 3, 4, 5, 6 \\ 0 & \text{if not} \end{cases}.$$

We can compute for example the probability of one defective fuse as

$$P(X = 1) = p_X(1) = \frac{C(10, 1)C(110, 5)}{C(120, 6)}$$

It will often be true that the distribution function of a discrete random variable will be piecewise constant; i.e., the real line can be partitioned into intervals on which F is constant. This is evidently the case in the preceding examples. However, some discrete random variables do not have such nice distribution functions; the following example is one of these.

Example: Let $\{r_1, r_2, \ldots, r_n, \ldots\}$ be the set of rational numbers in $[0, 1]$. Define

$$p(t) = \begin{cases} 1/2^n & \text{if } t = r_n \text{ for some } n \in \mathsf{N} \\ 0 & \text{if not} \end{cases}$$

Now p generates a distribution function $F(x) = \sum_{t \le x} p(t)$; F has a jump of height $1/2^n$ at r_n, so F is discontinuous at each rational number in $[0, 1]$. Since the particular ordering of the rational numbers in $[0, 1]$ is not explicitly given, it is impossible to sketch the graph of either p or F on $[0, 1]$.

For a discrete random variable, everything "happens" at points of discontinuity of the distribution function. We now pass to a type of random variable for which the distribution function has no points of discontinuity.

The distribution function for a discrete random variable is given in terms of a summation of values of a probability mass function. This produces a discontinuous distribution function. The continuous analogue of summation is, of course, integration. The functions which we shall integrate are piecewise continuous functions. A function $f: \mathsf{R} \to \mathsf{R}$ is *piecewise continuous* provided that R may be partitioned into a countable collection $\{I_n\}_1^\infty$ of intervals such that f is continuous on the interior of each interval.

Definition 2:   Let F_X be the distribution function of a random variable X on a p.m.s. (S, Σ, P). Suppose there exists a piecewise continuous function $f_X: \mathsf{R} \to \mathsf{R}$ such that F_X is given by the improper Riemann integral $F_X(x) = \int_{-\infty}^x f_X(t) \, dt$ for each $x \in \mathsf{R}$. Then f_X is called a *density* for X and X is said to be *absolutely continuous*. We shall sometimes say that F_X is absolutely continuous.

There are more general definitions of an absolutely continuous random variable in terms of more general densities or more general integrals (namely the Lebesgue integral). We have restricted ourselves to the

Riemann integral of piecewise continuous functions because these arise most often, but more importantly, because it is an easier job than introducing the advanced concepts. The student is advised, however, that in more advanced texts, the above definition may be broadened.

Example: Let

$$F(x) = \begin{cases} 0 & \text{if } x < 0 \\ x^2 & \text{if } 0 \leq x < 1; \\ 1 & \text{if } x \geq 1 \end{cases}$$

then a density for F is

$$f(x) = \begin{cases} 0 & \text{if } x < 0 \\ 2x & \text{if } 0 \leq x \leq 1. \\ 0 & \text{if } x > 1 \end{cases}$$

To prove that f is a density for F we integrate f and find

(1) for $x < 0$, $\displaystyle\int_{-\infty}^{x} f(t)\,dt = \int_{-\infty}^{x} 0\,dt = 0$

(2) for $0 \leq x \leq 1$, $\displaystyle\int_{-\infty}^{x} f(t)\,dt = \int_{-\infty}^{0} f(t)\,dt + \int_{1}^{x} f(t)\,dt$

$$= 0 + \int_{0}^{x} 2t\,dt = x^2$$

(3) for $x > 1$, $\displaystyle\int_{-\infty}^{x} f(t)\,dt = \int_{-\infty}^{1} f(t)\,dt + \int_{1}^{x} f(t)\,dt$

$$= 1 + 0 = 1.$$

Thus, $F(x) = \int_{-\infty}^{x} f(t)\,dt$ for each $x \in \mathbb{R}$.

The following theorem lists some important properties of distribution and density functions of an absolutely continuous random variable.

THEOREM 1: Let X be an absolutely continuous random variable on a p.m.s. (S, Σ, P) with distribution function F_X and density f_X. Then

(1) F_X is continuous.

(2) If f_X is continuous in some open interval containing x, then $f_X(x) = F_X'(x)$.

(3) If f_X is continuous in some open interval containing x, then $f_X(x) \geq 0$.

(4) $\int_{-\infty}^{\infty} f_X(t) \, dt = 1.$

(5)   For each $a \in \mathbb{R}$, $P(X = a) = 0.$

(6)   If $s_0 \in S$ and $\{s_0\} \in \Sigma$, then $P(\{s_0\}) = 0.$

Proof:

(1) $F_X(x) = \int_{-\infty}^{x} f_X(t) \, dt$ which is an improper integral so

$$\int_{-\infty}^{x} f_X(t) \, dt = \lim_{u \to x^-} \int_{-\infty}^{u} f_X(t) \, dt = \lim_{u \to x^-} F_X(u) = F_X(x^-).$$

We already know that $F_X(x) = F_X(x^+)$ so F_X is continuous at x.

(2)   Suppose f_X is continuous in an open interval containing x and let $\varepsilon > 0$; we shall show that there exists a $\delta > 0$ such that if $0 < |h| < \delta$, then

$$\left| \frac{F_X(x + h) - F_X(x)}{h} - f_X(x) \right| < \varepsilon.$$

Since f_X is continuous at x, there exists a $\delta > 0$ such that if $|x - t| < \delta$, then $|f_X(t) - f_X(x)| < \varepsilon$. Now let h be a number such that $0 < |h| < \delta$, then

$$\left| \frac{F_X(x + h) - F_X(x)}{h} - f_X(x) \right|$$

$$= \left| \frac{\int_{-\infty}^{x+h} f_X(t) \, dt - \int_{-\infty}^{x} f_X(t) \, dt}{h} - \frac{f_X(x) \cdot h}{h} \right|$$

$$= \frac{1}{|h|} \left| \int_{x}^{x+h} f_X(t) \, dt - \int_{x}^{x+h} f_X(x) \, dt \right|$$

$$\leq \frac{1}{|h|} \left| \int_{x}^{x+h} |f_X(t) - f_X(x)| \, dt \right|$$

$$< \frac{1}{|h|} \left| \int_{x}^{x+h} \varepsilon \, dt \right| = \frac{1}{|h|} \varepsilon |h| = \varepsilon.$$

(3)   By (2), $f_X(x) = F_X'(x)$; since F_X is non-decreasing, then its derivative (when it exists) is non-negative.

(4)
$$\int_{-\infty}^{\infty} f_X(t) = \lim_{x \to \infty} \int_{-\infty}^{x} f_X(t) \, dt = \lim_{x \to \infty} F_X(x) = 1$$

since F_X is a distribution function.

(5) $P(X = a) = F_X(a) - F_X(a^-) = 0$ since F_X is continuous.

(6) Let $X(s_0) = a$, if $\{s_0\} \in \Sigma$, then $\{s_0\} \subseteq \{s \in S \mid X(s) = a\}$ so $P(\{s_0\}) \leq P(X = a) = 0$. $\square$

In the example just preceding this theorem the density f_X was simply "pulled out of the air." This theorem shows that the choice is not completely arbitrary. If we know that a density f_X exists for a distribution function F_X, then the density is continuous except at, at most, a countable, separated set of points. We must have $f_X(x) = F_X'(x)$ at all except those points. The value of f_X at points of discontinuity *is* arbitrary.

If f is any non-negative function on R such that f is piecewise continuous and $\int_{-\infty}^{\infty} f(t) \, dt = 1$ then f determines a distribution function F by

$$F(x) = \int_{-\infty}^{x} f(t) \, dt.$$

The proof is left as an exercise.

The theorem above shows that the distribution function F_X for an absolutely continuous random variable X is continuous. Two comments should be made about this. First of all, it is *not* the random variable X which is continuous; there is no limit structure in S so there can be no concept of continuity for a function from S into R. Second, simply the continuity of F_X is not sufficient to ensure that X be absolutely continuous. (The "interested reader" may refer to any one of many texts on real analysis for a treatment of the "Cantor function" or "Lebesgue's singular function.") If F_X is continuous and has a finite derivative piecewise, however, then X *is* absolutely continuous. Statement (6) in the preceding theorem ensures that there are *no* absolutely continuous random variables defined on finite or countably infinite sample spaces.

Example: Let $n \in N$ be fixed and

$$F(x) = \begin{cases} 0 & \text{if } x < 0 \\ x^n & \text{if } 0 \leq x < 1 \\ 1 & \text{if } x \geq 1 \end{cases}.$$

F is continuous and has a finite derivative piecewise so it is the distribution function for an absolutely continuous random variable. We compute the density by taking F' wherever it exists:

$$f(x) = \begin{cases} 0 & \text{if } x \leq 0 \\ nx^{n-1} & \text{if } 0 < x < 1 \\ 0 & \text{if } x \geq 1 \end{cases}.$$

We have (more or less arbitrarily) assigned the density a value 0 wherever the derivative of F does not exist.

Example: Let $f(x) = 1/(\sqrt{2\pi})e^{-x^2/2}$ for $x \in \mathbb{R}$.

The function f is a density since $\int_{-\infty}^{\infty} e^{-x^2/2}\, dx = \sqrt{2\pi}$. It is called the *normal density* and the associated distribution

$$F(x) = \frac{1}{\sqrt{2\pi}} \int_{-\infty}^{x} e^{-t^2/2}\, dt$$

is called the *normal distribution*.

Example: The length of time a particular electronic component operates without failure is a random variable with distribution function

$$F(x) = \begin{cases} 0 & x < 0 \\ 1 - e^{-(x/100)} & x \geq 0 \end{cases}.$$

We observe that F is continuous on $\mathbb{R}$ and has a derivative except at $x = 0$, so F is the distribution function for an absolutely continuous random variable. A density is given by

$$f(x) = \begin{cases} 0 & x \leq 0 \\ (1/100)e^{-(x/100)} & x > 0 \end{cases}.$$

The probability that the component lasts for at least 100 time units is

$$P(X \geq 100) = 1 - F(100^-) = e^{-1} \doteq 0.3679.$$

This particular distribution has the following interesting property. Suppose that the component has been operating for at least T time units, then the probability that it operates for t more units is

$$P(X \geq T + t \mid X \geq T) = \frac{P(X \geq T + t)}{P(X \geq T)} = \frac{e^{-((T+t)/100)}}{e^{-(T/100)}}$$

$$= e^{-(t/100)} = P(X \geq t).$$

We know that every distribution function rises from 0 to 1. So far, we have discussed those which rise by jumps (distribution functions for discrete random variables) and those which rise smoothly (distribution functions for absolutely continuous random variables). We shall now consider those which do both.

If F_1 and F_2 are distribution functions and $0 \leq \lambda \leq 1$, then we know, from an exercise in the preceding section, that $\lambda F_1 + (1 - \lambda)F_2$ is also a distribution function. We can obtain a function that jumps at some points and rises smoothly at others by taking F_1 to be the distribution function of a discrete random variable and F_2 to be that of an absolutely continuous one.

Example: Let $\lambda = 1/2$ and F_1 and F_2 be given as follows:

$$F_1(x) = \begin{cases} 0 & \text{if } x < 0 \\ 1/3 & \text{if } 0 \leq x < 1 \\ 2/3 & \text{if } 1 \leq x < 2 \\ 1 & \text{if } x \geq 2 \end{cases} \qquad F_2(x) = \begin{cases} 0 & \text{if } x < 0 \\ 2x & \text{if } 0 \leq x < 1/2 \\ 1 & \text{if } x \geq 1/2 \end{cases}$$

If $F = \lambda F_1 + (1 - \lambda)F_2$ then F is given by

$$F(x) = \begin{cases} 0 & \text{if } x < 0 \\ 1/6 + x & \text{if } 0 \leq x < 1/2 \\ 1/6 + 1/2 = 2/3 & \text{if } 1/2 \leq x < 1 \\ 1/3 + 1/2 = 5/6 & \text{if } 1 \leq x < 2 \\ 1 & \text{if } x \geq 2 \end{cases}$$

The graphs are indicated in figure 5.8.

Example: A random variable is defined by the following process: A fair die is tossed; if 1, 2, 3, or 4 is obtained, then $X =$ that value; if 5 or 6 is

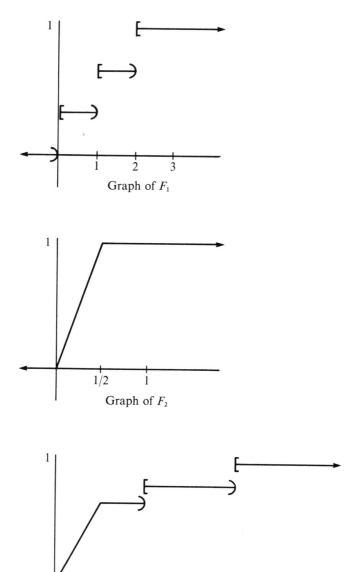

Graph of F_1

Graph of F_2

Graph of F

Figure 5.8

obtained, then a number is chosen from the interval $[1, 4]$ on the basis of the density

$$f(x) = \begin{cases} 1/3 & \text{if } 1 \leq x \leq 4 \\ 0 & \text{if not} \end{cases},$$

and $X =$ that value. Find F_X.

It may be helpful to consider the tree diagram in figure 5.9:

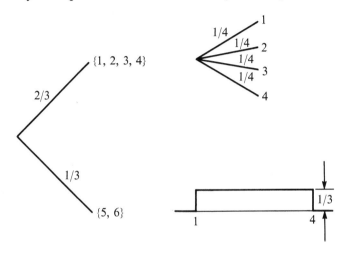

Figure 5.9

$$\begin{aligned} F_X(x) = P(X \leq x) &= P(X \leq x \text{ and } \{1, 2, 3, 4\}) \\ &\quad + P(X \leq x \text{ and } \{5, 6\}) \\ &= P(\{1, 2, 3, 4\}) \cdot P(X \leq x \mid \{1, 2, 3, 4\}) \\ &\quad + P(\{5, 6\}) \cdot P(X \leq x \mid \{5, 6\}) \\ &= (2/3)F_1(x) + (1/3)F_2(x), \end{aligned}$$

where F_1 is the distribution function for the discrete random variable with probability mass function

$$p_1(x) = \begin{cases} 1/4 & \text{if } x = 1, 2, 3, \text{ or } 4 \\ 0 & \text{if not} \end{cases}$$

and F_2 is the distribution function for the absolutely continuous random variable with density

$$f_2(x) = \begin{cases} 1/3 & \text{if } 1 \leq x \leq 4 \\ 0 & \text{if not} \end{cases}.$$

We compute

$$F_1(x) = \begin{cases} 0 & \text{if } x < 1 \\ 1/4 & \text{if } 1 \le x < 2 \\ 1/2 & \text{if } 2 \le x < 3 \\ 3/4 & \text{if } 3 \le x < 4 \\ 1 & \text{if } x \ge 4 \end{cases} \qquad F_2(x) = \begin{cases} 0 & \text{if } x < 1 \\ (x-1)/3 & \text{if } 1 \le x < 4 \\ 1 & \text{if } x \ge 4 \end{cases}$$

thus

$$F(x) = \begin{cases} 0 & \text{if } x < 1 \\ (2/3)(1/4) + (1/3)((x-1)/3) & \text{if } 1 \le x < 2 \\ (2/3)(1/2) + (1/3)((x-1)/3) & \text{if } 2 \le x < 3 \\ (2/3)(3/4) + (1/3)((x-1)/3) & \text{if } 3 \le x < 4 \\ 1 & \text{if } x \ge 4 \end{cases}$$

$$= \begin{cases} 0 & \text{if } x < 1 \\ 1/6 + (x-1)/9 & \text{if } 1 \le x < 2 \\ 1/3 + (x-1)/9 & \text{if } 2 \le x < 3 \\ 1/2 + (x-1)/9 & \text{if } 3 \le x < 4 \\ 1 & \text{if } x \ge 4 \end{cases}.$$

The graph of F is indicated in figure 5.10.

In the preceding example, we have three random variables: X, X_1, and X_2, and $F = \lambda F_1 + (1 - \lambda)F_2$ or $F_X = \lambda F_{X_1} + (1 - \lambda)F_{X_2}$. Note,

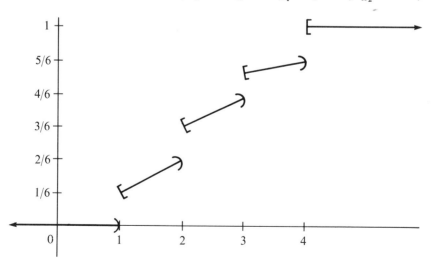

Figure 5.10

however, that it is *not* true that $X = \lambda X_1 + (1 - \lambda)X_2$. What is true is that in a sense X has a choice of how it will act; the probability is λ that it acts like X_1 and the probability is $1 - \lambda$ that it acts like X_2.

So far, we have been combining discrete and absolutely continuous distribution functions to produce another. If F is a distribution function which can be decomposed into a convex combination of a discrete and an absolutely continuous distribution function, then F is said to be *decomposable*. We shall also call the random variable X decomposable if its distribution function is decomposable. If F *is* decomposable, then it is uniquely decomposable and the components can be given in terms of F.

THEOREM 2: Suppose F is a decomposable distribution function; $F = \lambda F_1 + (1 - \lambda)F_2$ where F_1 is discrete and F_2 is absolutely continuous and $0 < \lambda < 1$. For each $x \in R$, let $p(x) = F(x) - F(x^-)$.

(1) The probability mass function p_1 for F_1 is given by $p_1(x) = (1/\lambda)p(x)$.

(2) $\lambda = \sum_{x \in R} p(x)$

(3) $F_1(x) = \dfrac{1}{\lambda} \sum_{t \le x} p(t)$

(4) $F_2(x) = (1/(1 - \lambda)) \cdot (F(x) - \lambda F_1(x))$.

Proof:

(1) Since F_2 is continuous, we have that

$$p(x) = F(x) - F(x^-)$$
$$= \lambda F_1(x) + (1 - \lambda)F_2(x) - \lambda F_1(x^-) - (1 - \lambda)F_2(x^-)$$
$$= \lambda[F_1(x) - F_1(x^-)] = \lambda p_1(x) \quad \text{from which (1) follows.}$$

(2) $\sum_{x \in R} p(x) = \sum_{x \in R} \lambda p_1(x) = \lambda \sum_{x \in R} p_1(x) = \lambda \cdot 1$ since F_1 is discrete.

(3) $F_1(x) = \sum_{t \le x} p_1(x) = \dfrac{1}{\lambda} \sum_{t \le x} p(x)$.

(4) This follows immediately from the fact that

$$F = \lambda F_1 + (1 - \lambda)F_2. \quad \square$$

In the preceding theorem, λ, F_1, and F_2 were derived solely from F. This seems to indicate that any distribution function can be decomposed. In a sense, this is correct: we can always decompose a distribution function F as $\lambda F_1 + (1 - \lambda)F_2$ where F_1 is discrete and F_2 is simply continuous. However, it may *not* be the case that F_2 is absolutely continuous. Of course if F_2 has a finite derivative piecewise, then F_2 is absolutely continuous.

Example: Decompose the distribution function

$$F(x) = \begin{cases} 0 & \text{if } x < 0 \\ x^2 & \text{if } 0 \leq x < 1/2 \\ 1/3 & \text{if } 1/2 \leq x < 1 \\ 1/2 + x/4 & \text{if } 1 \leq x < 2 \\ 1 & \text{if } x \geq 2 \end{cases}$$

The graph is given in figure 5.11.

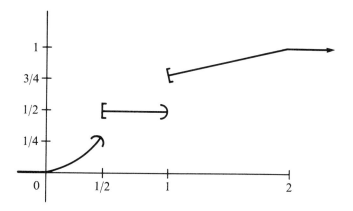

Figure 5.11

F has jumps at 1/2 and 1 of heights 1/12 and 5/12 respectively. Thus $\lambda = 1/12 + 5/12 = 1/2$ and

$$p_1(x) = 2 \begin{cases} 1/12 & \text{if } x = 1/2 \\ 5/12 & \text{if } x = 1 \\ 0 & \text{otherwise} \end{cases} = \begin{cases} 1/6 & \text{if } x = 1/2 \\ 5/6 & \text{if } x = 1 \\ 0 & \text{otherwise} \end{cases}.$$

We have that $F_1(x) = \sum_{t \leq x} p_1(x)$ so

$$F_1(x) = \begin{cases} 0 & \text{if } x < 1/2 \\ 1/6 & \text{if } 1/2 \leq x < 1 \\ 1 & \text{if } x \geq 1 \end{cases}.$$

$$F_2(x) = 2(F(x) - (1/2)F_1(x))$$

$$= 2 \begin{cases} 0 & \text{if } x < 0 \\ x^2 - 1/2(0) & \text{if } 0 \le x < 1/2 \\ 1/3 - 1/2(1/6) & \text{if } 1/2 \le x < 1 \\ 1/2 + x/4 - 1/2(1) & \text{if } 1 \le x < 2 \\ 1 - 1/2(1) & \text{if } x \ge 2 \end{cases}$$

$$= \begin{cases} 0 & \text{if } x < 0 \\ 2x^2 & \text{if } 0 \le x < 1/2 \\ 1/2 & \text{if } 1/2 \le x < 1. \\ x/2 & \text{if } 1 \le x < 2 \\ 1 & \text{if } x \ge 2 \end{cases}$$

The appropriate graphs are given in figure 5.12.

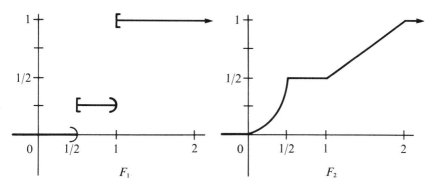

F_1 F_2

Figure 5.12

The density f_2 for F_2 is, of course,

$$f_2(x) = \begin{cases} 0 & \text{if } x \le 0 \\ 4x & \text{if } 0 < x < 1/2 \\ 0 & \text{if } 1/2 \le x \le 1. \\ 1/2 & \text{if } 1 < x < 2 \\ 0 & \text{if } x \ge 2 \end{cases}$$

Exercises:

1. Let (S, Σ, P) be a p.m.s. and let $A \in \Sigma$. Is χ_A a discrete, absolutely continuous, or decomposable r.v.?

2. For each of the following probability mass functions $p_X(x)$, find $F_X(x)$ and plot both. Find $P(-1 < X < 2)$ and $P(0 \le X \le 1)$.

(a) $p_X(x) = \begin{cases} 1/3 & \text{if } x = -1, 0, \text{ or } 1 \\ 0 & \text{otherwise} \end{cases}$

(b) $p_X(x) = \begin{cases} 1/n & \text{if } x = 1, 2, 3, \ldots, n \\ 0 & \text{otherwise} \end{cases}$

(c) $p_X(x) = \begin{cases} C(4, x)(1/3)^x(2/3)^{4-x} & \text{if } x = 0, 1, 2, 3, 4 \\ 0 & \text{otherwise} \end{cases}$
(Binomial probability mass function)

(d) $p_X(x) = \begin{cases} e^{-2}2^x/x! & \text{if } x = 0, 1, 2, 3, \ldots \\ 0 & \text{otherwise} \end{cases}$
(Poisson probability mass function)

3. For each of the following densities, find the constant c, the distribution function F, and plot the density and distribution function.

(a) $f(x) = \begin{cases} c & \text{if } a < x < b \\ 0 & \text{if not} \end{cases}$ a, b fixed, $a < b$ (uniform density)

(b) $f(x) = c/(1 + x^2)$ (Cauchy density)

(c) $f(x) = \begin{cases} c & \text{if } -1 < x < 1 \\ c/|x|^n & \text{if not} \end{cases}$ n fixed, $n > 1$

(d) $f(x) = \begin{cases} ce^{-x} & \text{if } x > 1 \\ 0 & \text{if not} \end{cases}$ (exponential density)

4. Find the density function for the distribution
$$F(x) = \begin{cases} 0 & \text{if } x < 0 \\ \sin x & \text{if } 0 \le x < \pi/2. \\ 1 & \text{if } x \ge \pi/2 \end{cases}$$

5. Convince yourself that each of the following distribution functions is decomposable. For each, find λ, the discrete part and probability mass function, and the absolutely continuous part and density.

(a) $F(x) = \begin{cases} 0 & x < -1 \\ (x + 1)^2/4 & -1 \le x < 0 \\ x + (1/2) & 0 \le x < 1/2 \\ 1 & x \ge 1/2 \end{cases}$

(b) $F(x) = \begin{cases} (1/3)e^x & x < 0 \\ 1 - (1/4)e^{-x} & x \ge 0 \end{cases}$

$$(c) \quad F(x) = \begin{cases} 0 & x < 0 \\ x/8 + 1/8 & 0 \le x < 1 \\ 1/4 & 1 \le x < 3 \\ 1 - 1/x^2 & x \ge 3 \end{cases}$$

6. Suppose $f: \mathsf{R} \to [0, \infty)$ is piecewise continuous and $\int_{-\infty}^{\infty} f(t) \, dt = 1$. Prove that f is a density for some absolutely continuous r.v.

7. If X is a r.v. on a p.m.s. (S, Σ, P) and the range $X(S)$ is countable, what can be said about X?

8. A lot of ten transistors is such that the probability of any one of them failing is $1/10$. The transistors are tested in succession until a defective one is found. Let X be the number of tests performed. Find p_X and indicate why X is discrete. Find $P(X \le 3)$.

9. A commuter figures that the length of time (in hours) that it takes him from his house to a railway station is a random variable with density as shown in figure 5.13. If he leaves home at 6:30 and the train leaves at 7:20, what is the probability that he will catch the train?

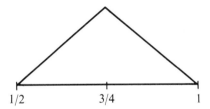

Figure 5.13

10. In the context of exercise 9: a traffic light is installed which catches him half the time and causes him to spend two minutes. Now answer the question in exercise 9.

5.4 Diversion into Analysis

In the preceding sections of this chapter, we defined random variables, classified them, and saw several examples. In this section we shall take a random variable and change it; we do this by composing it with another function. First, we need some definitions and results which properly belong to the areas of analysis and topology. The uninterested reader may simply move on to Theorem 1 and its corollaries.

Definition 1:   A subset O of R is said to be *open* provided that for each $a \in O$, there is an $\varepsilon > 0$ such that if $|x - a| < \varepsilon$, then $x \in O$, i.e.,

$$(a - \varepsilon, a + \varepsilon) \subseteq O.$$

Note that an open interval (c, d) is open since if $a \in (c, d)$, we may take $\varepsilon = \min (a - c, d - a)$. Then if $|x - a| < \varepsilon$, we have that $x \in (c, d)$. The empty set is open (vacuously) and R is open (let $\varepsilon = 1$ for any $a \in$ R, then $(a - 1, a + 1) \subseteq$ R). Other examples of open sets include those of the form $(-\infty, b)$ and (c, ∞). The set $[0, 1]$ is not open since we cannot surround 1 by an open interval completely contained in $[0, 1]$. Open sets can be combined by unions and intersections in certain ways to produce other open sets.

Proposition 1:   *If* $\{O_\alpha : \alpha \in \mathfrak{A}\}$ *is any collection of open sets, then* $\bigcup \{O_\alpha : \alpha \in \mathfrak{A}\}$ *is open. If* $\{O_1, \dots, O_n\}$ *is a finite collection of open sets then* $\bigcap_1^n O_i$ *is open.*

Proof: Let $O = \bigcup \{O_\alpha : \alpha \in \mathfrak{A}\}$; if $a \in O$, then $a \in O_\beta$ for some $\beta \in \mathfrak{A}$. Since this O_β is open, then there exists $\varepsilon > 0$ such that $(a - \varepsilon, a + \varepsilon) \subseteq O_\beta$. Now $(a - \varepsilon, a + \varepsilon) \subseteq O_\beta \subseteq O$ so O is open.   To prove the second statement of the proposition it is sufficient to prove that the intersection of any *two* open sets is open since the truth of the statement for any finite number follows easily by induction.   Let O_1 and O_2 be open and $a \in O_1 \cap O_1$. Then $a \in O_1$ and $a \in O_2$ so there exists $\varepsilon_1 > O$ and $\varepsilon_2 > 0$ such that $(a - \varepsilon_1, a + \varepsilon_1) \subseteq O_1$ and $(a - \varepsilon_2, a + \varepsilon_2) \subseteq O_2$.   Let $\varepsilon = \min (\varepsilon_1, \varepsilon_2)$; then $(a - \varepsilon, a + \varepsilon) \subseteq (a - \varepsilon_1, a + \varepsilon_1) \subseteq O_1$ and $(a - \varepsilon, a + \varepsilon) \subseteq (a - \varepsilon_2, a + \varepsilon_2) \subseteq O_2$. Thus $(a - \varepsilon, a + \varepsilon) \subseteq O_1 \cap O_2$ so $O_1 \cap O_2$ is open. $\square$

We saw in section 3 of chapter 2 that each open interval is a Borel set. The following proposition ensures that each open set is also a Borel set.

Proposition 2:   *Let O be an open subset of* R; *then O can be expressed as the union of a countable collection of open intervals.*

Proof: We shall use the fact that between any two real numbers, there exists a rational number.   This fact implies that each real number can be approximated arbitrarily closely by rational numbers—a statement not unreasonable since each real number has a decimal expansion and if we terminate that expansion at some point, we produce a rational number.

Let $a \in O$, then there exists $\varepsilon > 0$ such that $(a - \varepsilon, a + \varepsilon) \subseteq O$. Now there is a rational number b such that $a - \varepsilon/3 < b < a$. Let r be a positive rational number less than $2\varepsilon/3$ but greater than $a - b$ (this is possible since $a - b < \varepsilon/2$). Then $a \in (b - r, b + r) \subseteq (a - \varepsilon, a + \varepsilon) \subseteq O$. Thus O can be expressed as the union of (some) sets of the form $(b - r, b + r)$ where $b - r$ and $b + r$ are rational numbers. Now since the set of rationals is countable (see exercise 11, section 1.2) and so is the set of positive rationals, there can be only a countable number of such open intervals. Thus O is a union of some countable collection of open intervals. $\square$

Corollary 1:   Each open subset of R is a Borel set.

Proof: If O is open, then it is a countable union of open intervals. Each open interval is a Borel set and the collection of Borel sets is closed under countable union.

If $A \subseteq$ R and $f: A \to$ R, we recall that f is *continuous* at $a \in A$ provided that for each $\varepsilon > 0$, there is a $\delta > 0$ such that if $|x - a| < \delta$ and $x \in A$ then $|f(x) - f(a)| < \varepsilon$. We now give a characterization in terms of open sets.

Proposition 3:   *Let $f:$ R $\to$ R. Then these are equivalent:*

   (1)   *f is continuous at each member of* R

   (2)   *whenever O is open in* R, *then so is $f^{-1}(O)$.*

Proof: (1) $\Rightarrow$ (2): Let O be open in R and let $U = f^{-1}(O)$. If $a \in U$, then $f(a) \in O$ so there exists an $\varepsilon > 0$ such that $(f(a) - \varepsilon, f(a) + \varepsilon) \subseteq O$. Now since f is continuous at a, there exists $\delta > 0$ such that if $|x - a| < \delta$, then $|f(x) - f(a)| < \varepsilon$. Thus if $x \in (a - \delta, a + \delta)$, then

$$f(x) \in (f(a) - \varepsilon, f(a) + \varepsilon) \subseteq O.$$

This implies that $(a - \delta, a + \delta) \subseteq f^{-1}(O) = U$ so U is open.

   (2) $\Rightarrow$ (1): Let $a \in$ R and $\varepsilon > 0$; then $O = (f(a) - \varepsilon, f(a) + \varepsilon)$ is an open set (every open interval is an open set). Thus $f^{-1}(O)$ is open and $a \in f^{-1}(O)$; so there exists a $\delta > 0$ such that if $|x - a| < \delta$, then $x \in f^{-1}(O)$. This implies that $f(x) \in O = (f(a) - \varepsilon, f(a) + \varepsilon)$ and so if $|x - a| < \delta$, then $|f(x) - f(a)| < \varepsilon$. $\square$

The interested reader is invited to invent a proposition similar to Proposition 3 but concerning continuity on an arbitrary subset of R.

The preceding proposition implies that continuous functions "back up" open sets into open sets. Of perhaps more importance in probability is the fact (see Proposition 4, below) that continuous functions back up Borel sets into Borel sets.

Definition 2:   Let $f: \mathsf{R} \to \mathsf{R}$, f is a *Borel function* provided that whenever B is a Borel subset of R, then so is $f^{-1}(B)$.

Proposition 4:   *Every continuous function $f: \mathsf{R} \to \mathsf{R}$ is a Borel function.*

Proof: We use the same "trick" as that in Theorem 1 of section 1 of this chapter. Let $\mathscr{M} = \{A \subseteq \mathsf{R} \mid f^{-1}(A) \text{ is a Borel set}\}$; then by exactly the same steps as before, $\mathscr{M}$ is a σ-algebra. Now, if I is an open interval, then it is an open set and so $f^{-1}(I)$ is open. But each open set is also a Borel set, so $f^{-1}(I)$ is a Borel set and we have shown that $I \in \mathscr{M}$. Thus $\mathscr{M}$ is a σ-algebra which includes each open interval so it includes each Borel set. $\square$

The proposition assures us of the existence of many Borel functions but there are Borel functions which are not continuous.

Example: If B is any Borel set, then

$$\chi_B = \begin{cases} 1 & \text{on } B \\ 0 & \text{on } \mathsf{R}\backslash B \end{cases}$$

is a Borel function.

Let A be any Borel set in R, we must show that $\chi_B^{-1}(A)$ is a Borel set. Now

if $0 \notin A$ and $1 \notin A$, then $\chi_B^{-1}(A) = \varnothing$;

if $0 \notin A$ and $1 \in A$, then $\chi_B^{-1}(A) = B$;

if $0 \in A$ and $1 \notin A$, then $\chi_B^{-1}(A) = \mathsf{R}\backslash B$;

if $0 \in A$ and $1 \in A$, then $\chi_B^{-1}(A) = \mathsf{R}$;

so in any case $\chi_B^{-1}(A)$ is a Borel set.

In chapter 2, we defined the Borel subsets of R; in a similar fashion, we can define the Borel subsets of R^n for each $n \in N$.

Definition 3:   Let $n \in \mathsf{N}$, then the *Borel sets of* R^n, the collection of which is denoted by $\mathscr{B}(\mathsf{R}^n)$, are the members of the smallest σ-algebra containing all sets of the form $(-\infty, a_1] \times (-\infty, a_2) \times \cdots \times (-\infty, a_n]$ where $a_1, a_2, \ldots, a_n$ are real numbers.

As before, there are many ways in which $\mathscr{B}(\mathsf{R}^n)$ may be generated and most common subsets of R^n are Borel sets.

We could have defined the concept of open subset of R^n in a similar fashion to those of R (working with the product of n open intervals) and exactly the same proposition as Proposition 1 would hold. Continuity of functions from R^n to R^m has an open set characterization exactly as that in Proposition 3. The concept of a Borel function from R^n to R^m can also be defined as above, and each continuous function is a Borel function. We summarize this in the following proposition.

Proposition 5: *Let $f: \mathsf{R}^n \to \mathsf{R}^m$ be continuous, then f is a Borel function.*

By this time, the reader may be asking himself what all this has to do with random variables. The following theorem answers that question.

THEOREM 1: Let (S, Σ, P) be a p.m.s. and $X_1, X_2, \ldots, X_n$ be random variables. Let $f: \mathsf{R}^n \to \mathsf{R}$ be a Borel function. Then the function

$$Y(s) = f(X_1(s), X_2(s), \ldots, X_n(s))$$

is a random variable.

Lemma 1.1: Let $Z: S \to \mathsf{R}^n$ by $Z(s) = (X_1(s), X_2(s), \ldots, X_n(s))$. Then for each Borel set B of R^n, $Z^{-1}(B) \in \Sigma$ (Z might be called an R^n-*valued random variable*).

Proof: If $a_1, a_2, \ldots, a_n$ are real numbers, then

$$Z^{-1}((-\infty, a_1] \times (-\infty, a_2] \times \cdots \times (-\infty, a_n])$$
$$= \{s \in S \mid X_1(s) \le a_1, X_2(s) \le a_2, \ldots, X_n(s) \le a_n\}$$
$$= X_1^{-1}((-\infty, a_1]) \cap X_2^{-1}((-\infty, a_2]) \cap \cdots \cap X_n^{-1}((-\infty, a_n]).$$

Since each X_i is a r.v., this intersection is in Σ. If $\mathscr{M}$ is the collection of all subsets of R^n such that $A \in \mathscr{M}$ if and only if $Z^{-1}(A) \in \Sigma$ then $\mathscr{M}$ is a σ-algebra, and we have just proved that $\mathscr{M}$ contains all sets of the form $(-\infty, a_1] \times \cdots \times (-\infty, a_n]$. Thus $\mathscr{M}$ contains all Borel sets of R^n. $\quad\square$

Proof of Theorem 1: Let $a \in \mathsf{R}$, then $(-\infty, a]$ is a Borel set in R, hence $f^{-1}((-\infty, a])$ is a Borel set in R^n. Now $Z^{-1}(f^{-1}((-\infty, a])) \in \Sigma$ by Lemma 1.1, but $Z^{-1}(f^{-1}((-\infty, a])) = Y^{-1}((-\infty, a])$. $\quad\square$

Corollary 1.1:   Let (S, Σ, P) be a p.m.s. and $X_1, \ldots, X_n$ be random variables.   Let $f: \mathbb{R}^n \to \mathbb{R}$ be continuous.   Then the function $Y(s) = f(X_1(s), \ldots, X_n(s))$ is a r.v.

Corollary 1.2:   Let X be a r.v. on a p.m.s. (S, Σ, P) and let $g: \mathbb{R} \to \mathbb{R}$ be a Borel function.   Let $Y: S \to \mathbb{R}$ by $Y(s) = g(X(s))$.   Then Y is a r.v.

Corollary 1.3:   Let X and Y be random variables on a p.m.s. (S, Σ, P); let $a \in \mathbb{R}$.   Then

     (1) $X + Y$ is a r.v.

     (2) aX is a r.v.

     (3) $X - Y$ is a r.v.

     (4) $X \cdot Y$ is a r.v.

Proof: The functions of addition, multiplication by a constant, subtraction, and multiplication are continuous. $\square$

Example: Suppose that X is a r.v. and has a uniform density on $[0, 1]$. Let $Y = 1/(1 + X^2)$.

Let $g: \mathbb{R} \to \mathbb{R}$ by $g(t) = 1/(1 + t^2)$, then g is continuous so g is a Borel function.   Now $Y(s) = g(X(s)) = 1/(1 + [X(s)]^2)$ is a r.v. by Corollary 1.2.   Let us find a density for Y.   First note that since $P(0 \le X \le 1) = 1$, then $P(1 \le 1 + X^2 \le 2) = 1$ and thus $P(1 \ge 1/(1 + X^2) \ge 1/2) = 1$; so for $t \notin [1/2, 1], f_Y(t) = 0$.   Now for $t \in [1/2, 1]$,

$$F_Y(t) = P(Y \le t) = P(1/(1 + X^2) \le t) = P(1 + X^2 \ge 1/t)$$
$$= P(X^2 \ge 1/t - 1)$$
$$= P(X \ge \sqrt{1/t - 1})$$
$$= 1 - F_X(\sqrt{1/t - 1}).$$

Thus

$$f_Y(t) = \frac{d}{dt}(F_Y(t)) = \frac{d}{dt}\left(1 - F_X\left(\sqrt{\frac{1}{t} - 1}\right)\right)$$

$$= -f_X\left(\sqrt{\frac{1}{t} - 1}\right)\left(\frac{1}{2}\frac{1}{\sqrt{1/t - 1}}\left(-\frac{1}{t^2}\right)\right)$$

$$= \begin{cases} \sqrt{t}/(2t^2\sqrt{1 - t}) & \text{if } 1/2 \le t \le 1 \\ 0 & \text{if not} \end{cases}.$$

Exercises:

1. Prove Proposition 1 for R^n rather than R.

2. A subset A of R is *closed* if $\mathsf{R}\backslash A$ is open. Prove that the union of two closed sets is closed. Prove that the intersection of any family of closed sets is closed. Prove that each closed set is a Borel set. Show that $f: \mathsf{R} \to \mathsf{R}$ is continuous if and only if for each closed set A of R, $f^{-1}(A)$ is closed.

3. Let $A \subseteq \mathsf{R}$ and $f: A \to \mathsf{R}$. Show that these are equivalent:
   (1) f is continuous at each point of A.
   (2) If O is any open set in R, then there is an open set U in R such that $f^{-1}(O) = U \cap A$.

4. Suppose $f: \mathsf{R} \to \mathsf{R}$. Show that f is a Borel function if and only if for each *open* set B of R, $f^{-1}(B)$ is a Borel set.

5. Prove that if f and g are Borel functions, from R to R, then so are $f + g$, $af(a \in \mathsf{R})$, $f - g$, $f \cdot g$, and the composition $f \circ g(x) = f(g(x))$. (Hint: use Corollary 1.2 with an appropriate p.m.s.)

6. Suppose I is an interval in R and $f: \mathsf{R} \to \mathsf{R}$ such that f is continuous on I and if $x \notin I$, then $f(x) = 0$. Prove that f is a Borel function.

7. Suppose $f: \mathsf{R} \to \mathsf{R}$ is continuous except at a finite number of points; prove that f is a Borel function.

8. If X is a r.v. on a p.m.s. (S, Σ, P), then show that each of the following is a r.v. and find the distribution functions in terms of F_X.
   (a) $Y(s) = aX(s) + b$.
   (b) $Y(s) = [X(s)]^2$.
   (c) $Y(x) = \sin X(s)$.
   (d) $Y(s) = e^{X(s)}$
   (e) $Y(s) = [X(s)]^3$.
   (f) $Y(s) = \begin{cases} 0 & \text{if } X(s) < 0 \\ 1 & \text{if } X(s) \geq 0 \end{cases}$.

9. X is a r.v. with density $f_X(t) = (1/\sqrt{2\pi})e^{-(t^2/2)}$. Find a density for $Y = X^2$.

10. X is a r.v. with the Cauchy density and
$$Y(s) = \begin{cases} X(s) & \text{if } -1 \leq X(s) \leq 1 \\ 0 & \text{if not} \end{cases}.$$
    Show that Y is a r.v. and find a density for Y.

6

Expectation

6.1 Definitions

Suppose an experiment has as outcomes, the numbers $a_1, a_2, \ldots, a_n$ with probabilities $p_1, p_2, \ldots, p_n$ respectively. If the experiment is repeated a large number of times (say N times), then we expect that a_1 will occur on approximately $p_1 \cdot N$ of the times, that a_2 will occur on approximately $p_2 \cdot N$ of the times, and so on. The average outcome for the N trials is, of course, the sum of the outcomes on the N trials, divided by N. If we group together all of the a_1's, all of the a_2's, etc., then we expect the average to be approximately

$$\frac{(p_1 N)a_1 + (p_2 N)a_2 + \cdots + (p_n N)a_n}{N} = \sum_{i=1}^{n} p_i a_i.$$

As N increases, we expect the fraction of occurrences of a_1 to get closer to p_1 so the average should get closer and closer to $\sum_1^n p_i a_i$ (we shall see that, in a sense, this is true). The discussion above motivates the following definition.

Definition 1:   Let X be a discrete random variable with probability mass function p_X. We define the *expectation* (or *mean*) of X to be

$$E(X) = \sum_{\text{all } x} x p_X(x)$$

provided that $\sum_{\text{all } x} |x| p_X(x)$ is finite.

  Since $p_X(x) = 0$ for all but a countable number of x's, then $x p_X(x) = 0$ for all but a countable number of x's; by assuming that $\sum_{\text{all } x} |x| p_X(x)$ is finite, we conclude that the order in which we add up the terms $x p_X(x)$ is irrelevant.

Example: Let $a_1, a_2, \ldots, a_n$ be distinct real numbers and

$$p_X(x) = \begin{cases} 1/n & \text{if } x = a_i \text{ for some } i \\ 0 & \text{if not} \end{cases}.$$

p_X is a probability mass function and we compute

$$E(X) = \sum_{\text{all } x} x p_X(x) = \sum_{i=1}^n a_i \cdot \frac{1}{n} = \frac{1}{n} \sum_{i=1}^n a_i;$$

this is our usual concept of (arithmetic) mean or average. Note that $\sum_{\text{all } x} |x| p_X(x)$ is finite since only a *finite* number of $p_X(x)$'s are non-zero.

Example: A fair die is tossed until a 3 is obtained.

If X is the random variable giving the number of times until 3, then the probability mass function for X is

$$p_X(x) = \begin{cases} (1/6)(5/6)^{n-1} & \text{if } x = n \in \mathsf{N} \\ 0 & \text{if not} \end{cases}.$$

This is derived from the fact that $X = n$ when the first $n - 1$ tosses result in non-3's and the last toss in a 3. Now p_X is a probability mass function since

$$\sum_{x \in R} p_X(x) = \sum_{n=1}^{\infty} \frac{1}{6} \left(\frac{5}{6}\right)^{n-1} = \frac{1}{6} \sum_{n=1}^{\infty} \left(\frac{5}{6}\right)^{n-1} = \frac{1}{6} \left(\frac{1}{1 - 5/6}\right) = 1.$$

By the definition,

$$E(X) = \sum_{x \in R} x p_X(x) = \sum_{n=1}^{\infty} n \frac{1}{6} \left(\frac{5}{6}\right)^{n-1} = \frac{1}{6} \sum_{n=1}^{\infty} n \left(\frac{5}{6}\right)^{n-1}.$$

To sum this final series, we recall that if $\sum_{n=0}^{\infty} a_n x^n$ is a power series, then within its radius of convergence, the derivative is $\sum_{n=1}^{\infty} n a_n x^{n-1}$. Thus $\sum_{n=1}^{\infty} n(x)^{n-1}$ is the derivative of $\sum_{n=0}^{\infty} x^n$ for $|x| < 1$; but $\sum_{n=0}^{\infty} x^n = 1/(1 - x)$ and the derivative of $1/(1 - x)$ is $1/(1 - x)^2$. Thus

$$E(X) = \frac{1}{6} \sum_{n=1}^{\infty} n \left(\frac{5}{6}\right)^{n-1} = \frac{1}{6} \frac{1}{(1 - 5/6)^2} = (1/6)(36) = 6.$$

Definition 2: Let X be an absolutely continuous r.v. with density f_X, then we define the *expectation (or mean) of* X to be $E(X) = \int_{-\infty}^{\infty} x f_X(x) \, dx$ provided $\int_{-\infty}^{\infty} |x| f_X(x) \, dx < \infty$.

The reader may notice a great deal of similarity between the formulae for $E(X)$ and the definition of the first moment of a system as he saw it in physics; in fact, the concepts are exactly the same from a certain mathematical standpoint. The only difference is that we require the total mass of our system to be 1.

Example: X is a r.v. with density

$$f_X(x) = \begin{cases} 1/(b - a) & \text{if } a < x < b \\ 0 & \text{if not} \end{cases}.$$

X is said to be uniformly distributed on the interval (a, b).

$$E(X) = \int_{-\infty}^{\infty} x(f_X(x)) \, dx$$

$$= \int_{a}^{b} x \left(\frac{1}{b - a}\right) dx = \frac{1}{b - a} \left(\frac{b^2}{2} - \frac{a^2}{2}\right) = \frac{b + a}{2}$$

which is, of course, half way between a and b. The criterion that $\int_{-\infty}^{\infty} |x| f_X(x)\, dx < \infty$ is met since $|x|/(b - a)$ is bounded on (a, b).

Example: Let X have density $f_X(x) = (1/\pi)(1/(1 + x^2))$. f_X is a density since

$$\frac{1}{\pi} \int_{-\infty}^{\infty} \frac{1}{1 + x^2}\, dx = \frac{1}{\pi} \left[\lim_{x \to \infty} \tan^{-1}(x) - \tan^{-1}(-x) \right]$$

$$= \frac{1}{\pi} \left(\frac{\pi}{2} + \frac{\pi}{2} \right) = 1.$$

Now

$$\int_{-\infty}^{\infty} \frac{|x|}{1 + x^2}\, dx = 2 \int_0^{\infty} \frac{x}{1 + x^2}\, dx$$

which is infinite. Thus X has no expectation.

Example: Let X have density $f_X(x) = (1/\sqrt{2\pi})e^{-(x^2/2)}$.

One usually proves by polar coordinates and double integrals that $\int_{-\infty}^{\infty} e^{-x^2}\, dx = \sqrt{\pi}$ and from this it follows that f_X is a density. We have that

$$\frac{1}{\sqrt{2\pi}} \int_{-\infty}^{\infty} |x| e^{-(x^2/2)} = 2 \frac{1}{\sqrt{2\pi}} \int_0^{\infty} x e^{-(x^2/2)}\, dx \quad \text{since} \quad |x| e^{-(x^2/2)}$$

is symmetric around $x = 0$. Now by a change of variables $u = x^2/2$,

$$\int_0^{\infty} x e^{-(x^2/2)}\, dx = \int_0^{\infty} e^{-u}\, du = e^{-u}\big|_0^{\infty} = 1,$$

so

$$\int_{-\infty}^{\infty} |x| f_X(x)\, dx = \frac{2}{\sqrt{2\pi}}$$

which is finite. Thus X has an expectation and we compute easily that

$$E(X) = \frac{1}{\sqrt{2\pi}} \int_{-\infty}^{\infty} x e^{-(x^2/2)}\, dx = 0.$$

In previous sections, we have considered decomposable random variables—those whose distribution functions are convex combinations of discrete and absolutely continuous distribution functions. One primary reason for singling out these particular random variables is that one may define a reasonable notion of expectation for them.

Definition 3:   Let X be a r.v. whose distribution function F_X can be expressed as $F_X = \lambda F_1 + (1 - \lambda)F_2$ where F_1 is a discrete distribution function with probability mass function p_1 and F_2 is an absolutely continuous distribution function with density f_2 and $0 < \lambda < 1$. Then we define the *expectation (or mean) of X* to be

$$E(X) = \lambda \sum_{x \in R} x p_1(X) + (1 - \lambda) \int_{-\infty}^{\infty} x f_2(x)\, dx$$

provided

$$\sum_{x \in R} |x|\, p_1(x) < \infty \quad \text{and} \quad \int_{-\infty}^{\infty} |x| f_2(x)\, dx < \infty.$$

Example: Let X be a r.v. with distribution function

$$F_X(x) = \begin{cases} 0 & x < 0 \\ x/2 & 0 \le x < 1 \\ 1 & x \ge 1 \end{cases}.$$

Then

$$\lambda = 1/2, \quad p_1(x) = \begin{cases} 1 & \text{if } x = 1 \\ 0 & \text{if not} \end{cases}, \quad \text{and} \quad f_2(x) = \begin{cases} 0 & x \le 0 \\ 1 & 0 < x < 1 \\ 0 & x \ge 1 \end{cases}.$$

Thus

$$E(X) = \frac{1}{2}(1 \cdot 1) + \frac{1}{2}\left(\int_0^1 x\, dx \right) = \frac{1}{2} + \frac{1}{2}\left(\frac{1}{2}\right) = \frac{3}{4}.$$

Let us recall that if $p_X(x)$ is the jump of F at x, then $p_X(x) = \lambda p_1(x)$ so the first summand in the definition of $E(X)$ when X is decomposable could be written $\sum_{x \in R} x p_X(x)$. Similarly, if F_X is piecewise differentiable, then at all but a countable number of points $F_X' = (1 - \lambda)f_2$. Thus, the second summand can sometimes be written $\int_{-\infty}^{\infty} x F_X'(x)\, dx$. We must note, however, that p_X is *not* a probability mass function and F_X' is *not* a density unless X can be decomposed in a trivial way ($\lambda = 0$ or $\lambda = 1$).

In succeeding sections we shall see some properties of expectations and more examples.

Example: A r.v. X has a distribution function,

$$F_X(t) = \begin{cases} 0 & t < 0 \\ (1/4)\sqrt{t} & 0 \le t < 1 \\ 1/2 & 1 \le t < 2 \\ t/4 & 2 \le t < 3 \\ 1 & t \ge 3 \end{cases}.$$

Now F is piecewise differentiable so we may use the simpler formula for $E(X)$ mentioned in the preceding paragraph.

$$p_X(t) = \begin{cases} 1/4 & \text{if } t = 1 \text{ or } 3 \\ 0 & \text{if not} \end{cases}$$

$$F_X'(t) = \begin{cases} 1/8(1/\sqrt{t}) & \text{if } 0 < t < 1 \\ 1/2 & \text{if } 2 < t < 3 \\ 0 & \text{if not} \end{cases}.$$

$$E(X) = \sum_{t \in \mathbb{R}} t p_X(t) + \int_{-\infty}^{\infty} t F_X'(t)\, dt$$

$$= 1\left(\frac{1}{4}\right) + 3\left(\frac{1}{4}\right) + \int_0^1 \frac{1}{8}\sqrt{t}\, dt + \int_2^3 \frac{t}{2}\, dt$$

$$= 1 + (1/8)((2/3)t^{3/2}|_0^1) + (1/2)(t^2/2)|_2^3$$

$$= 1 + 1/12 + 5/4 = 28/12 = 7/3.$$

Exercises:

1. Find an example of a probability mass function p_X such that $E(X)$ does not exist.

2. In exercise 2 of section 5.3, several examples of probability mass functions are given. Find the expectation of the r.v. associated with each.

3. If

$$p_X(x) = \begin{cases} (1 - p)p^{n-1} & \text{if } x = n \in \mathbb{N} \\ 0 & \text{if not} \end{cases},$$

show that p_X is a probability mass function and find $E(X)$ (p fixed).

4. In exercise 3 of section 5.3, several examples of density functions are given. For each, find whether $E(X)$ exists and if it does compute it.

*5. Suppose that X is an absolutely continuous r.v. and is symmetric around $a \in \mathbb{R}$ (i.e. for each $x \geq 0$, $P(X < a - x) = P(X \geq a + x)$). Prove that f_X is symmetric about a (i.e. for each $x > 0$, $f_X(a - x) = f_X(a + x)$) whenever f_X is continuous and find $E(X)$ under the assumption that it exists.

6. In exercise 5 of section 5.3, several examples of decomposable distribution functions are given. For each, find the expectation of the r.v. associated with it if this expectation exists.

7. A die is tossed either four times or until a 3 appears. Find the expected number of tosses.

8. The length of time that an electronic component operates without failure is a r.v. X with distribution function

$$F_X(t) = \begin{cases} 0 & t < 0 \\ 1 - e^{-(t/T)} & t \geq 0 \end{cases}$$

where T is a constant. Find $E(X)$.

9. Let (S, Σ, P) be a p.m.s. and $A \in \Sigma$. Define

$$\chi_A(s) = \begin{cases} 1 & \text{if } s \in A \\ 0 & \text{if } s \notin A \end{cases}.$$

Prove that $E(\chi_A) = P(A)$.

10. Let (S, Σ, P) be a p.m.s. and $A, B \in \Sigma$, with A and B independent and $P(A) = 1/2 = P(B)$. Define

$$X: S \to \mathbb{R} \text{ by } X(s) = \begin{cases} 0 & \text{if } s \in (A \cup B)^c \\ 2 & \text{if } s \in A \cap B \\ 1 & \text{if } s \in (A \cup B) \backslash A \cap B \end{cases}.$$

Find $E(X)$.

6.2 Expectation of a Function with Respect to a Random Variable

Let X be a random variable and $g: \mathbb{R} \to \mathbb{R}$ be a function. In this section we consider an expectation of g relative to the random variable X. Instead of giving separate definitions relative to discrete, absolutely continuous, and then decomposable random variables, we combine all the definitions and give them under the decomposable case.

Definition 1: Let X be a r.v. whose distribution function F can be decomposed as $\lambda F_1 + (1 - \lambda)F_2$ where F_1 is discrete with probability mass function p_1 and F_2 is absolutely continuous with density f_2 and $0 \le \lambda \le 1$. Let $g: R \to R$ be Riemann integrable in case $\lambda \ne 1$ and be such that both $\lambda \sum_{x \in R} |g(x)| p_1(x)$ and $(1 - \lambda) \int_{-\infty}^{\infty} |g(x)| f_2(x)\, dx$ are finite. The expectation of g relative to X is

$$E_X(g) = \lambda \sum_{x \in R} g(x) p_1(x) + (1 - \lambda) \int_{-\infty}^{\infty} g(x) f_2(x)\, dx.$$

Example: Let X be a discrete r.v. with probability mass function

$$p_X(x) = \begin{cases} 1/6 & x = 0 \\ 1/3 & x = 1 \\ 1/2 & x = 3 \\ 0 & \text{otherwise} \end{cases} \quad . \quad \text{Let } g: R \to R \text{ by } g(x) = x^2 + 1.$$

Then

$$E_X(g) = \sum_{x \in R} g(x) p_X(x)$$

$$= 1 \cdot \frac{1}{6} + 2 \cdot \frac{1}{3} + 10 \cdot \frac{1}{2} = \frac{1 + 4 + 30}{6} = \frac{35}{6}.$$

Example: X has density

$$f(x) = \begin{cases} 1/(b - a) & \text{if } a < x < b \\ 0 & \text{otherwise} \end{cases} \quad ; \quad g(x) = x^2.$$

Then

$$E_X(g) = \int_{-\infty}^{\infty} g(x) f(x)\, dx = \int_a^b x^2 \frac{1}{b - a}\, dx$$

$$= \frac{1}{b - a} \frac{x^3}{3} \Big|_a^b = \frac{1}{3} \frac{b^3 - a^3}{b - a} = \frac{1}{3}(b^2 + ab + a^2).$$

Example: Let

$$F_X(x) = \begin{cases} 0 & \text{if } x < 0 \\ x/2 & \text{if } 0 \le x < 1 \\ 1 & \text{if } x \ge 1 \end{cases}.$$

Then X is decomposable with $\lambda = 1/2$,

$$p_1(x) = \begin{cases} 1 & \text{if } x = 1 \\ 0 & \text{if not} \end{cases},$$

and

$$f_2(x) = \begin{cases} 0 & \text{if } x \leq 0 \\ 1 & \text{if } 0 < x < 1 \\ 0 & \text{if } x \geq 1 \end{cases}.$$

Let $g(x) = x^3$. Then

$$E_X(g) = \frac{1}{2}(1^3 \cdot 1) + \frac{1}{2}\int_0^1 x^3 \cdot 1 \, dx = \frac{1}{2} + \frac{1}{2}\left(\frac{1}{4}\right) = \frac{5}{8}.$$

We note directly from the definition that $E_X(x) = E(X)$, where by $E_X(x)$ we mean the expectation of the identity function $g(x) = x$ relative to X. We note also that if A is an interval in R, that $E_X(\chi_A) = P(X \in A)$ (see exercise 1, section 6.2). Some important facts concerning expectation of functions with respect to a r.v. are given in the following theorem.

THEOREM 1:   Let X be a decomposable r.v.

(1)   If $g(x) = c$ for each $x \in$ R, then $E_X(g) = c$.

(2)   If g_1 and g_2 are functions from R into R and both $E_X(g_1)$ and $E_X(g_2)$ exist, then $E_X(g_1 + g_2)$ exists and $E_X(g_1 + g_2) = E_X(g_1) + E_X(g_2)$.

(3)   If $E_X(g)$ exists and $c \in$ R, then $E_X(cg)$ exists and $E_X(cg) = c \cdot E_X(g)$.

(4)   If $g: \text{R} \to [0, \infty)$ and $E_X(g)$ exists, then $E_X(g) \geq 0$.

(5)   If g_1 and g_2 are functions from R into R such that $g_1(x) \leq g_2(x)$ for each $x \in$ R and both $E_X(g_1)$ and $E_X(g_2)$ exist then $E_X(g_1) \leq E_X(g_2)$.

(6)   If $E_X(g)$ exists then $E_X(|g|)$ exists and $|E_X(g)| \leq E_X(|g|)$.

Proof: Let $F_X = \lambda F_1 + (1 - \lambda)F_2$ where F_1 is discrete with probability mass function p_1, F_2 is absolutely continuous with density f_2, and $0 \leq \lambda \leq 1$. We shall assume $f_2(x) \geq 0$ for each $x \in$ R.

(1) $E_X(g) = \lambda \sum_{x \in R} g(x) p_1(x) + (1 - \lambda) \int_{-\infty}^{\infty} g(x) f_2(x)\, dx$

$= \lambda \sum_{x \in R} c p_1(x) + (1 - \lambda) \int_{-\infty}^{\infty} c f_2(x)\, dx$

$= c \left(\lambda \sum_{x \in R} p_1(x) + (1 - \lambda) \int_{-\infty}^{\infty} f_2(x)\, dx \right).$

Since p_1 is a probability mass function, $\sum_{x \in R} p_1(x) = 1$ and since f_2 is a density, $\int_{-\infty}^{\infty} f_2(x)\, dx = 1$. Thus

$$E_X(g) = c(\lambda + (1 - \lambda)) = c.$$

(2) Since $|g_1(x) + g_2(x)| \leq |g_1(x)| + |g_2(x)|$ for each $x \in R$, then

$\lambda \sum_{x \in R} |g_1(x) + g_2(x)| p_1(x)$

$+ (1 - \lambda) \int_{-\infty}^{\infty} |g_1(x) + g_2(x)| f_2(x)\, dx$

$\leq \lambda \left(\sum_{x \in R} |g_1(x)| p_1(x) + \sum_{x \in R} |g_2(x)| p_1(x) \right)$

$+ (1 - \lambda) \left(\int_{-\infty}^{\infty} |g_1(x)| f_2(x)\, dx + \int_{-\infty}^{\infty} |g_2(x)| f_2(x)\, dx \right)$

$= E_X(|g_1|) + E_X(|g_2|) < \infty.$

Thus $E_X(g_1 + g_2)$ exists. Now

$E_X(g_1 + g_2) = \lambda \sum_{x \in R} (g_1(x) + g_2(x)) p_1(x)$

$+ (1 - \lambda) \int_{-\infty}^{\infty} (g_1(x) + g_2(x) f_2(x)\, dx$

which can be broken up as before to yield $E_X(g_1) + E_X(g_2)$.

(3) We shall leave this as an exercise.

(4) If $g(x) \geq 0$ for each $x \in R$, then $g(x)p_1(x) \geq 0$ and $g(x)f_2(x) \geq 0$
for each $x \in R$. Thus

$$E_X(g) = \lambda \left(\sum_{x \in R} g(x)p_1(x) \right) + (1 - \lambda) \left(\int_{-\infty}^{\infty} g(x)f_2(x)\, dx \right) \geq 0.$$

(5) If $g_1(x) \leq g_2(x)$, then $g_2(x) - g_1(x) \geq 0$ for each $x \in R$. Thus
$E_X(g_2 - g_1) \geq 0$ by (4). But by use of (1) and (3), we find that
$E_X(g_2 - g_1) = E_X(g_2) - E_X(g_1)$. Thus $E_X(g_2) - E_X(g_1) \geq 0$
from which (5) follows.

(6) For each $x \in R$, $-|g(x)| \leq g(x) \leq |g(x)|$. Thus $E_X(-|g|) \leq$
$E_X(g) \leq E_X(|g|)$ by (5).   But $E_X(-|g|) = -E_X(|g|)$ so
$-E_X(|g|) \leq E_X(g) \leq E_X(|g|)$ from which it follows that
$|E_X(g)| \leq E_X(|g|)$. □

Recall from section 5.4 that if $X: S \to R$ is a r.v. and $g: R \to R$ is a Borel
function, then $Y: S \to R$ by $Y(s) = g(X(s))$ is a r.v. also. In most of the
cases with which we shall be dealing, whenever X is decomposable, Y will
be decomposable also and in this case,

$$E(Y) = E_X(g).$$

Example: Suppose X is a discrete r.v. and $g(t) = t^2$.

$$Y(s) = (X(s))^2$$

and

$$p_Y(t) = P(Y = t) = P(X^2 = t)$$

$$= \begin{cases} 0 & \text{if } t < 0 \\ P(X = 0) & \text{if } t = 0 \\ P(X = \sqrt{t}) + P(X = -\sqrt{t}) & \text{if } t > 0 \end{cases}$$

$$= \begin{cases} 0 & \text{if } t < 0. \\ p_X(0) & \text{if } t = 0 \\ p_X(\sqrt{t}) + p_X(-\sqrt{t}) & \text{if } t > 0 \end{cases}$$

Thus $\sum_{t \in R} p_Y(t) = \sum_{t > 0} [p_X(\sqrt{t}) + p_X(-\sqrt{t})] + p_X(0) = \sum_{t \in R} p_X(t) = 1$
so Y is discrete also.

$$E(Y) = \sum_{t \in R} t p_Y(t) = \sum_{t > 0} [t p_X(\sqrt{t}) + t p_X(-\sqrt{t})] + p_X(0)$$

$$= \sum_{u > 0} u^2 p_X(u) + \sum_{u > 0} u^2 p_X(u) t p_X(0)$$

$$= \sum_{u \in R} u^2 p_X(u) = E_X(u^2).$$

Example: Let X be an absolutely continuous r.v. and $g(t) = t^2$.

$$F_Y(t) = P(Y \le t) = P(X^2 \le t)$$

$$= \begin{cases} 0 & \text{if } t < 0 \\ P(-\sqrt{t} \le X \le \sqrt{t}) & \text{if } t \ge 0 \end{cases}$$

$$= \begin{cases} 0 & \text{if } t < 0 \\ F_X(\sqrt{t}) - F_X(-\sqrt{t}^-) & \text{if } t \ge 0 \end{cases}$$

$$= \begin{cases} 0 & \text{if } t < 0 \quad \text{since } F_X \text{ is continuous.} \\ F_X(\sqrt{t}) - F_X(-\sqrt{t}) & \text{if } t \ge 0 \end{cases}$$

Thus,

$$f_Y(t) = \begin{cases} 0 & t < 0 \\ F_Y'(t) = \dfrac{d}{dt}(F_X(\sqrt{t}) - F_X(-\sqrt{t})) & t \ge 0 \end{cases}$$

$$= \begin{cases} 0 & t < 0 \\ f_X(\sqrt{t}) \cdot (1/2)(1/\sqrt{t}) - f_X(-\sqrt{t}) \cdot (-(1/2)(1/\sqrt{t})) & t \ge 0 \end{cases}$$

$$= \begin{cases} 0 & t < 0 \\ (1/2)(1/\sqrt{t})(f_X(\sqrt{t}) + f_X(-\sqrt{t})) & t \ge 0 \end{cases}$$

$$E(Y) = \int_{-\infty}^{\infty} t f_Y(t) \, dt = \frac{1}{2} \int_0^{\infty} \sqrt{t}(f_X(\sqrt{t}) + f_X(-\sqrt{t})) \, dt$$

$$= \frac{1}{2} \int_0^{\infty} \sqrt{t} f_X(\sqrt{t}) \, dt + \frac{1}{2} \int_0^{\infty} \sqrt{t} f_X(-\sqrt{t}) \, dt.$$

By a change of variable $u = \sqrt{t}$ in the first integral and $u = -\sqrt{t}$ in the second, we see that the sum of the two integrals is

$$\int_0^{\infty} u^2 f_X(u) \, du + \int_{-\infty}^0 u^2 f_X(u) \, du = \int_{-\infty}^{\infty} u^2 f_X(u) \, du$$

which is $E_X(u^2)$ or $E_X(t^2)$.

Example: Let X be decomposable and $g(t) = t^2$.

Suppose $F_X = \lambda F_1 + (1 - \lambda)F_2$ as usual. Then let X_1 be a r.v. such that $F_{X_1} = F_1$ and X_2 be a r.v. such that $F_{X_2} = F_2$. If $Y = X^2$, then

$$F_Y(t) = P(Y \le t) = P(X^2 \le t)$$

$$= \begin{cases} 0 & t < 0 \\ F_X(\sqrt{t}) - F_X(-\sqrt{t^-}) & t \ge 0 \end{cases}$$

$$= \begin{cases} 0 & t < 0 \\ \lambda F_1(\sqrt{t}) + (1 - \lambda)F_2(\sqrt{t}) - \lambda F_1(-\sqrt{t^-}) \\ \qquad\qquad\qquad\qquad - (1 - \lambda)F_2(-\sqrt{t^-}) & t \ge 0 \end{cases}$$

$$= \begin{cases} 0 & t < 0 \\ \lambda(F_1(\sqrt{t}) - F_1(-\sqrt{t^-})) \\ \qquad\qquad + (1 - \lambda)(F_2(\sqrt{t}) - F_2(-\sqrt{t^-})) & t \ge 0 \end{cases}$$

$$= \lambda F_{Y_1}(t) + (1 - \lambda)F_{Y_2}(t)$$

where $Y_1 = (X_1)^2$ and $Y_2 = (X_2)^2$ by the preceding examples. It is now a simple task to show that

$$E(Y) = E_X(t^2).$$

We are faced with problems in the above assumption ($E_X(g) = E(Y)$) only when Y is not decomposable. Then it is necessary to bring into play a generalization of the Riemann integral in order to define a concept of expectation for Y. When this generalization is made, it is still the case that $E_X(g) = E(Y)$.

Example: A traffic light is on a one-minute cycle with the first thirty seconds green, the next ten seconds yellow, and the last twenty seconds red. What is the expected length of time spent at the traffic light?

Fix a point 100 feet before the light and let X be a random variable giving the time that a motorist passes that point. Because of the periodicity of the light, we shall assume that X is uniformly distributed over the interval $[0, 60]$. Now for each time $t \in [0, 60]$, there is a particular probability that the motorist will stop at the light if he passes the point at time t. Let $g(t)$ be this probability; then g might have a graph as given in figure 6.1.

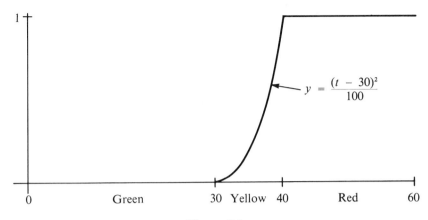

Figure 6.1

Let Y be the random variable giving the length of time the motorist spends at the light. Then $Y = g(X) \cdot (60 - X)$ so

$$E(Y) = E_X(g(t) \cdot (60 - t)) = \int_0^{60} \frac{g(t)(60 - t)}{60}\, dt$$

$$= \int_0^{30} 0\, dt + \int_{30}^{40} \frac{(t - 30)^2(60 - t)}{6000}\, dt + \int_{40}^{60} \frac{60 - t}{60}\, dt$$

$$= 0 - \int_{30}^{40} \frac{(t - 30)^2(t - 30 - 30)}{6000}\, dt + \int_{40}^{60} 1 - \frac{t}{60}\, dt$$

$$= 30 \int_{30}^{40} \frac{(t - 30)^2}{6000}\, dt - \int_{30}^{40} \frac{(t - 30)^3}{6000}\, dt + \left(t - \frac{t^2}{120} \Big|_{40}^{60} \right)$$

$$= 55/12.$$

Exercises:

1. Show that if X is a decomposable r.v. and A is an interval in R, then $E_X(\chi_A) = P(X \in A)$; compare this with exercise 9 of the preceding section.

2. Show that if X is a discrete r.v. and A is a Borel set, then $E_X(\chi_A) = P(X \in A)$.

*3. Show that if X is a decomposable r.v. and $E_X(g)$ exists and $c \in$ R, then $E_X(cg) = cE_X(g)$.

4. Finish the example in this section to show that if X is decomposable, then $E(X^2) = E_X(x^2)$.

5. Show that if X is decomposable, then $E(X^3) = E_X(x^3)$. You may want to treat the discrete and absolutely continuous cases first.

6. Let X be a decomposable r.v. Show that $E_X(\sin x)$ exists.

7. Let X be a decomposable r.v. and let g be a non-negative function such that $E_X(g)$ exists. Show that if f is improper Riemann integrable and $|f(x)| \leq g(x)$ for each $x \in R$, then $E_X(f)$ exists.

8. Let X be a r.v. with normal density $f(t) = (1/\sqrt{2\pi})e^{-(t^2/2)}$. Find $E(X^2)$.

*9. Let X be a r.v. on a p.m.s. and let $Y(s) = aX(s) + b$. Show directly that $E(Y) = aE(X) + b$.

6.3   Moments of a Random Variable

One reason for introducing the notion of a distribution function for a r.v. X was that real valued functions of a real variable are often easier to analyze than real valued functions on a sample space. We continue the analysis in the present section, investigating such questions as "how spread out are the values of X?" and "is X symmetric in some sense?"

Definition 1:   Let X be a decomposable r.v. For each $n \in N$, we define the nth *moment of X* to be $E_X(x^n)$ whenever it exists.
    We recall from the previous section that

$$E_X(x^n) = E(X^n)$$

and we shall write the latter in most cases.

Example: Let X be a discrete r.v. with probability mass function

$$p_X(x) = \begin{cases} 1/6 & \text{if } x = 0, 1, 2, 3, 4, \text{ or } 5 \\ 0 & \text{if not} \end{cases}.$$

The first three moments are

$$E(X) = \sum_{x \in R} x p_X(x)$$

$$= \sum_{n=0}^{5} n \frac{1}{6} = \frac{1}{6}(0 + 1 + 2 + 3 + 4 + 5) = \frac{15}{6} = \frac{5}{2},$$

$$E(X^2) = \sum_{x \in R} x^2 p_X(x)$$

$$= \sum_{n=0}^{5} n^2 \frac{1}{6} = \frac{1}{6}(0 + 1 + 4 + 9 + 16 + 25) = \frac{55}{6},$$

$$E(X^3) = \sum_{x \in R} x^3 p_X(x)$$

$$= \sum_{n=0}^{5} n^3 \frac{1}{6} = \frac{1}{6}(0 + 1 + 8 + 27 + 64 + 125) = \frac{225}{6}.$$

Example: Let X be an absolutely continuous r.v. with density

$$f(x) = \begin{cases} 1/(b - a) & \text{if } a \le x \le b \\ 0 & \text{if not} \end{cases}.$$

The first three moments are

$$E(X) = \int_{-\infty}^{\infty} x f(x) \, dx = \frac{1}{b - a} \int_{a}^{b} x \, dx$$

$$= \frac{x^2}{2(b - a)} \bigg|_a^b = \frac{b + a}{2}$$

$$E(X^2) = \int_{-\infty}^{\infty} x^2 f(x) \, dx = \frac{1}{b - a} \int_{a}^{b} x^2 \, dx$$

$$= \frac{x^3}{3(a - b)} \bigg|_a^b = \frac{b^2 + ba + a^2}{3}$$

$$E(X^3) = \int_{-\infty}^{\infty} x^3 f(x) \, dx = \frac{1}{b - a} \int_{a}^{b} x^3 \, dx$$

$$= \frac{x^4}{4(b - a)} \bigg|_a^b = \frac{b^3 + b^2 a + ba^2 + a^3}{4}.$$

We shall introduce shortly a function which often facilitates computation of moments but first we state a theorem that we shall find necessary in the investigation of that function. The proof of the theorem does not involve a significant number of new concepts but is rather long, requiring several lemmas. We shall refer the reader who is interested in the details to R. G. Bartle's *The Elements of Real Analysis*, published in 1964 by John Wiley, pp. 359 ff.

DOMINATED CONVERGENCE THEOREM: Let $\{f_n\}$ be a sequence of (improper) Riemann integrable functions on R and f and g be (improper) Riemann integrable functions such that

(1)   for each $x \in$ R, $f_n(x) \to f(x)$

(2)   for each $x \in$ R, and each $n \in$ N, $|f_n(x)| \le g(x)$.

Then $\int_{-\infty}^{\infty} f_n(x)\, dx \to \int_{-\infty}^{\infty} f(x)\, dx$

$$\left(\text{i.e. } \lim_{n \to \infty} \int_{-\infty}^{\infty} f_n(x)\, dx = \int_{-\infty}^{\infty} \lim_{n \to \infty} f_n(x)\, dx \right).$$

Definition 2:   Let X be a decomposable r.v. We define the *moment generating function* of X to be $M_X(t) = E_X(e^{tX})$ wherever this expectation exists.

Suppose X is absolutely continuous and $M_X(t)$ exists for all t in the interval $[-T, T]$. Then for each $t \in [-T, T]$, and each $x \in$ R,

$$\sum_0^N \frac{(tx)^n}{n!} \to e^{tx} \quad \text{and} \quad \left| \sum_0^N \frac{(tx)^n}{n!} \right| \le \sum_1^N \frac{|tx|^n}{n!} \le e^{|tx|} \le e^{T|x|}.$$

Thus by the Dominated Convergence Theorem we have

$$\int_{-\infty}^{\infty} \sum_1^N \frac{(tx)^n}{n!} f_X(x)\, dx \to \int_{-\infty}^{\infty} e^{(tx)} t_X(x)\, dx = M_X(t).$$

But

$$\int_{-\infty}^{\infty} \sum_1^N \frac{(tx)^n}{n!} f_X(x)\, dx = \sum_1^N \int_{-\infty}^{\infty} \frac{(tx)^n}{n!} f_X(x)\, dx,$$

thus

$$M_X(t) = \sum_1^{\infty} \int_{-\infty}^{\infty} \frac{(tx)^n}{n!} f_X(x)\, dx = \sum_1^{\infty} \left(\int_{-\infty}^{\infty} x^n f_X(x)\, dx \right) \frac{t^n}{n!}.$$

We shall use this fact in the following theorem.

THEOREM 1: Let X be decomposable and M_X exist in $[-T, T]$ ($T > 0$). Then for each $n \in N$, $M_X^{[n]}(0) = E(X^n)$ ($M_X^{[n]}$ denotes the nth derivative of M_X).

Proof: Let $F_X = \lambda F_1 + (1 - \lambda)F_2$ as usual. Then

$$\sum_{x \in R} e^{tx} p_1(x) = \sum_{x \in R}\left(\sum_{n=0}^{\infty} \frac{(tx)}{n!} p_1(x)\right),$$

and since this sum converges absolutely we can rearrange terms to give

$$\sum_{n=0}^{\infty}\left(\sum_{x \in R} \frac{(tx)^n}{n!} p_1(x)\right) = \sum_{n=0}^{\infty} (x^n p_1(x)) \frac{t^n}{n!}.$$

Therefore, using the expression derived above for an absolutely continuous r.v., we have

$$M_X(t) = \lambda \sum_{x \in R} e^{tx} p_1(x) + (1 - \lambda) \int_{-\infty}^{\infty} e^{tx} f_2(x)\, dx$$

$$= \lambda \sum_{n=0}^{\infty}\left(\sum_{x \in R} x^n p_1(x)\right) \frac{t^n}{n!} + (1 - \lambda) \sum_{n=0}^{\infty}\left(\int_{-\infty}^{\infty} x^n f_2(x)\, dx\right) \frac{t^n}{n!}$$

$$= \sum_{n=0}^{\infty} (E(X^n)) \frac{t^n}{n!}.$$

Now the latter series is a power series in t and converges inside $[-T, T]$, so we may take the derivative of M_X at 0, term by term. This produces $M_X(0) = E(X)^0 = 1$, $M_X'(0) = E(X^1)$, etc. $\square$

In practice, for the random variables for which it is easy to compute the moments directly, the moment generating function method is often significantly more complicated. However, for many random variables for which direct computation is difficult, the moment generating function helps tremendously.

Example: Let X be a r.v. with probability mass function given as follows where n is fixed and $0 < p < 1$.

$$p(x) = \begin{cases} C(n, x)\, p^x(1 - p)^{n-x} & \text{if } x = 0, 1, 2, \ldots, n \\ 0 & \text{if not} \end{cases}.$$

We note that p is a probability mass function since

$$\sum_{x \in R} p(x) = \sum_{k=0}^{n} C(n, k) p^{k}(1 - p)^{n-k} = (p + (1 - p))^{n} = 1^{n} = 1.$$

We shall compute $E(X)$ and $E(X^2)$ directly and by use of the moment generating function.

Direct Computation:

$$E(X) = \sum_{k=0}^{n} kC(n, k) p^{k}(1 - p)^{n-k}$$

$$= \sum_{k=1}^{n} k \frac{n!}{k! \, (n - k)!} \, p^{k}(1 - p)^{n-k}$$

$$= np \sum_{k=1}^{n} \left[\frac{(n - 1)!}{(k - 1)! \, ((n - 1) - (k - 1))!} \right.$$

$$\left. \times \, p^{k-1}(1 - p)^{(n-1)-(k-1)} \right]$$

$$= np(p + (1 - p))^{n-1} = np.$$

$$E(X^2) = \sum_{k=0}^{n} k^2 C(n, k) p^{k}(1 - p)^{n-k}.$$

Note that $k^2 = k(k - 1) + k$; so

$$E(X^2) = \sum_{k=0}^{n} k(k - 1)C(n, k) p^{k}(1 - p)^{n-k}$$

$$+ \sum_{k=0}^{n} kC(n, k) p^{k}(1 - p)^{n-k}$$

$$= \sum_{k=2}^{n} \left[\frac{n!}{(k - 2)! \, ((n - 2) - (k - 2))!} \right.$$

$$\left. \times \, p^{k}(1 - p)^{(n-2)-(k-2)} \right] + E(X)$$

$$= n(n - 1)p^2 \sum_{k=2}^{n} \frac{(n - 2)!}{(k - 2)! \, ((n - 2) - (k - 2))!}$$

$$\times \, p^{k-2}(1 - p)^{(n-2)-(k-2)} + E(X)$$

$$= n(n - 1)p^2(p + (1 - p))^{n-2} + np$$

$$= np((n - 1)p + 1).$$

Moment Generating Function:

Since $p_X = 0$ except at a finite number of points, then $M_X(t)$ exists for each $t \in \mathsf{R}$. Now

$$M_X(t) = E(e^{tx}) = \sum_{k=0}^{n} e^{tk} C(n, k) p^k (1 - p)^{n-k}$$

$$= \sum_{k=0}^{n} C(n, k)(e^t p)^k (1 - p)^{n-k}$$

$$= (e^t p + 1 - p)^n.$$

Thus $M_X'(t) = n(e^t p + 1 - p)^{n-1}(e^t p)$ so $E(X) = M_X'(0) = n(p + 1 - p)^{n-1}(p) = np$.
$M_X''(t) = n(n - 1)(e^t p + 1 - p)^{n-2}e^t p e^t p + n(e^t p + 1 - p)^{n-1}e^t p$ so
$E(X^2) = M_X''(0) = n(n - 1)(p + 1 - p)^{n-2}pp + n(p + 1 - p)^{n-1}p = n(n - 1)p^2 + np$.

Example: X is absolutely continuous with density

$$f(x) = \begin{cases} 0 & \text{if } x < 0 \\ e^{-x} & \text{if } x \geq 0 \end{cases}.$$

We shall compute only $E(X)$ by the direct method but both $E(X)$ and $E(X^2)$ by the moment generating function.

Direct Computation:

$$E(X) = \int_{-\infty}^{\infty} x f(x)\, dx = \int_{0}^{\infty} x e^{-x}\, dx = \lim_{a \to \infty} \int_{0}^{a} x e^{-x}\, dx.$$

Using integration by parts, we find that

$$\int_{0}^{a} x e^{-x}\, dx = (1 - e^{-a}) - a e^{-a}$$

which has limit 1 as $a \to \infty$, thus $E(X) = 1$.

Moment Generating Function:

$$M_X(t) = \int_{0}^{\infty} e^{tx} e^{-x}\, dx = \int_{0}^{\infty} e^{(t-1)x}\, dx$$

$$= \frac{1}{t - 1} e^{(t-1)x} \Big|_{0}^{\infty} = \frac{1}{t - 1} (0 - 1)$$

$$= -(1/(t - 1)) = 1/(1 - t) = (1 - t)^{-1} \quad \text{for} \quad -1 < t < 1.$$

Now $M_X(t)$ exists for $t \in [-(1/2), 1/2]$ so the moments may be generated.

$$M_X'(t) = (-1)(1 - t)^{-2}(-1) = (1 - t)^{-2}$$
$$M_X''(t) = (-2)(1 - t)^{-3}(-1) = 2(1 - t)^{-3}.$$

Thus $E(X) = M_X'(0) = (1 - 0)^{-2} = 1$

$$E(X^2) = M_X''(0) = 2(1 - 0)^{-2} = 2.$$

Now that we have computed some moments, let's see how moments can be used to tell us something about the random variable. First, we seek a measure of the dispersion of the values of X. In a sense, the expectation $E(X)$ is the center of the values of X so it is reasonable to measure the dispersion from $E(X)$ (for another reason, see exercise 4 of this section). As a first attempt at a measure of dispersion we might consider $E(X - E(X))$, that is, the expected distance from the expectation or mean. However, if we let $m = E(X)$, and consider "m" to be the constantly m function, then we have $E(X - m) = E_X(x - m) = E_X(x) - E_X(m) = m - m = 0$; so this is not a reasonable measure of dispersion. As a second (and more successful) attempt we shall make sure that values on the left of m and values on the right of m do not cancel out as before. There are several ways in which this may be accomplished and the simplest is to investigate $E(|X - m|)$. This quantity is called the *mean deviation of X*. This measure is a reasonable measure of dispersion but analysis involving the mean deviation often can be rather complicated (to integrate a function involving an absolute value one must, in general, integrate the function over two or more subintervals). Another way to protect against "cancelling out" is simply to raise $(X - m)$ to an even power, so that we investigate $E((X - m)^n)$ where n is even. This bears a resemblance to the nth moment, and, in fact, we make the following definition.

Definition 2:   Let X be a decomposable r.v. such that $E(X) = m$ exists, and let $n \in N$. We define the *nth central moment* of X to be $E_X((x - m)^n) = E((X - m)^n)$ whenever that expectation exists.

We have shown above that the first central moment is always 0. The second central moment $E((X - m)^2)$ is a measure of dispersion and is called the *variance* of X (denoted Var (X) or σ_X^2). We define the *standard deviation* of X by $\sigma_X = \sqrt{\text{Var }(X)}$.

The third central moment $E((X - m)^3)$ is not a measure of dispersion since if X is symmetric around $a \in R$, then the third moment is 0 (see exercise 5).

The fourth central moment is again a measure of dispersion since we are raising $(X - m)$ to an even power. Here, however, if $|X - m| < 1$ then $(X - m)^4$ is much less than 1 and if $|X - m| > 1$, then $(X - m)^4$ is much greater than 1; so those values of X a great distance away from m are weighted much more heavily than are those only a short distance away.

If the moments of X have been computed, then the central moments may be computed directly from them. For example, we have the following theorem.

THEOREM 2: If $E(X)$ and $E(X^2)$ exist, then

$$\text{Var}\,(X) = E((X - m)^2) = E(X^2) - [E(X)]^2.$$

Proof: Let $m = E(X)$. Then

$$
\begin{aligned}
E((X - m)^2) &= E(X^2 - 2Xm + m^2) = E_X(x^2 - 2xm + m^2) \\
&= E_X(x^2) - 2mE_X(x) + m^2 \\
&= E_X(x^2) - 2m \cdot m + m^2 = E(X^2) - [E(X)]^2. \quad \square
\end{aligned}
$$

Example: Let X be a r.v. and $Y = aX + b$. Find Var (Y) in terms of Var (X).

$$
\begin{aligned}
\text{Var}\,(Y) &= E(Y^2) - [E(Y)]^2 \\
&= E(a^2X^2 + 2abX + b^2) - [aE(X) + b]^2 \\
&= a^2E(X^2) + 2abE(X) + b^2 - a^2[E(X)]^2 - 2abE(X) - b^2 \\
&= a^2(E(X^2) - [E(X)]^2) \\
&= a^2\,\text{Var}\,(X).
\end{aligned}
$$

In particular, if $Y = (X - m)/\sigma_X = (1/\sigma_X)X - m/\sigma_X$, then $E(Y) = 0$ and Var $(Y) = 1$; this is called the *standardization* of X.

As an example of how the variance of X may be used to give information about the dispersion of X we have:

CHEBYSHEV'S INEQUALITY: Let X be a random variable with expectation m and variance σ^2 (thus $\sigma = $ standard deviation). Then for each $h > 0$, $P(m - h\sigma < X \le m + h\sigma) \ge 1 - 1/h^2$.

Proof: Let $A = (m - h\sigma, m + h\sigma]$.   Then $\sigma^2 = E_X((x - m)^2) \geq E_X((x - m)^2 \chi_{R/A}(x))$ since $(x - m)^2 \geq (x - m)^2 \chi_{R/A}(x)$ for each $x \in$ R. But if $x \in$ R\\$A$, then either $x \leq m - h\sigma$ or $x > m + h\sigma$ and in either case $(x - m)^2 \geq h^2\sigma^2$. Thus $\sigma^2 \geq h^2\sigma^2 E_X(\chi_{R/A}) = \sigma^2 h^2 P(X \in$ R\\$A)$ and so $1/h^2 \geq P(X \in$ R\\$A)$ from which it follows that $P(X \in A) \geq 1 - 1/h^2$.

$\square$

For $0 < h \leq 1$, Chebyshev's Inequality tells us nothing, but for $h > 1$, useful information is produced; we have, for example, that

$$P(m - 3\sigma < X \leq m + 3\sigma) \geq 1 - 1/9 = 8/9.$$

Chebyshev's Inequality is quite useful for dealing with general random variables; but for many random variables, Chebyshev's Inequality gives a very conservative estimate.

Example: Let X have a density

$$f(x) = \begin{cases} 1 & \text{if } 0 \leq x \leq 1 \\ 0 & \text{if not} \end{cases}$$

Now $m = E(X) = 1/2$ and $\sigma^2 = E(X^2) - [E(X)]^2 = 1/3 - 1/4 = 1/12$ (by the example at the first of this section); so $\sigma = 1/\sqrt{12} = 1/(2\sqrt{3})$. Now

$$P(m - \sigma h < X \leq m + \sigma h)$$
$$= P(1/2 - h/(2\sqrt{3}) < X < 1/2 + h/(2\sqrt{3})) = h/\sqrt{3}$$

for $h \leq \sqrt{3}$ and $P(m - \sigma h < X \leq m + \sigma h) = 1$ for $h > \sqrt{3}$. For $h = (2\sqrt{3})/3$, $p(m - \sigma h < X \leq m + \sigma h) = 2/3$ while Chebyshev's Inequality says only that

$$p(m - \sigma h < X \leq m + \sigma h) \geq 1 - 9/12 = 1/4.$$

Exercises:

1.   Compute directly first and second moments of the r.v. which has density

$$p(x) = \begin{cases} e^{-\lambda}\lambda^x/x! & \text{if } x = 0, 1, 2, \ldots \\ 0 & \text{if not} \end{cases} \quad \text{where } \lambda > 0 \text{ is fixed.}$$

2. Compute the moment generating function for the r.v. corresponding to the probability mass function above and then compute the first and second moments.

3. Compute the moment generating functions for the random variables described below. Then compute the first and second moments.
   (a) X has probability mass function

$$p(x) = \begin{cases} (1 - p)p^x & \text{if } x = 0, 1, 2, \dots \\ 0 & \text{if not} \end{cases}$$

where $0 < p < 1$ is fixed.
   (b) X has density $f(x) = (1/\sqrt{2\pi})e^{-(x^2/2)}$.
   (c) X has density

$$f(x) = \begin{cases} 1/(b - a) & \text{if } a \le x \le b \\ 0 & \text{if not} \end{cases}$$

   (d) X has distribution function

$$F_X(x) = \begin{cases} 0 & \text{if } x \le 0 \\ x/2 & \text{if } 0 \le x < 1 \\ 3/4 & \text{if } 1 \le x < 2 \\ 1 & \text{if } x \ge 2 \end{cases}$$

   (e) X has distribution function

$$F_X(x) = \begin{cases} 0 & \text{if } x < 0 \\ 1 - e^{-(x/T)} & \text{if } x \ge 0 \end{cases} \quad T \text{ constant.}$$

4. Let X be a r.v. such that Var (X) exists. Define $f: \mathrm{R} \to \mathrm{R}$ by $f(a) = E((X - a)^2)$. Show that f has a minimum at $a = E(X)$ ($E(X)$ is therefore said to be a least squares estimate).

*5. Suppose X is a r.v. such that $E((X - m)^3)$ exists and suppose X has the property that if $x \ge 0$, then $P(X \le m - x) = P(X \ge m + x)$; then X is said to be symmetric around m. Show that the third central moment is 0.

6. Suppose X is a r.v. such that the first three moments exist. Express $E((X - m)^3)$ in terms of these moments (see Theorem 2).

7. The temperature of an autoclave during a particular chemical reaction is observed to have a mean of 200° (F) and a variance of 100° (F^2). At a randomly selected time, the temperature is measured. What can be said about the probability that the temperature is higher than 250° F?

*8. Suppose that X is a r.v. such that $E(X) = m$ and Var $(X) = 0$. Prove that

$$P(X = m) = 1.$$

(i.e. X is almost constant).

7

Jointly Distributed Random Variables

7.1 Joint Distribution Function

In the preceding chapters, we have been concerned mainly with one random variable at a time. Often, there is more than one r.v. in which we are interested; we could treat these random variables individually, but they are sometimes related in ways which can be useful. We shall treat the case of two random variables in detail and then indicate how our treatment can be generalized.

Definition 1: Let X and Y be random variables on a probability measure space (S, Σ, P). The *joint distribution* for X and Y is the function $F_{X,Y}: \mathsf{R}^2 \to \mathsf{R}$ defined by

$$F_{X,Y}(x, y) = P(\{s \in S \mid X(s) \le x\} \cap \{s \in S \mid Y(s) \le y\}).$$

Note that since X and Y are random variables, then both

$$\{s \in S \mid X(s) \le x\} \quad \text{and} \quad \{s \in S \mid Y(s) \le y\}$$

are events and thus their intersection is an event. We shall usually abbreviate the probability of this event by $P(X \le x, Y \le y)$.

Example: Let $S = \{a, b\}$, $\Sigma = \mathscr{P}(S)$, and $P(\{a\}) = P(\{b\}) = 1/2$ (this, of course, determines P on all of Σ). Let $X \colon S \to \mathsf{R}$ by $X(a) = 0$, $X(b) = 1$ and let $Y \colon S \to \mathsf{R}$ by $Y(a) = 1$, $Y(b) = -1$.

The pair $(X(s), Y(s))$ forms a two-dimensional vector so we have a R^2-valued function on S. In particular, $(X(a), Y(a)) = (0, 1)$; $(X(b), Y(b)) = (1, -1)$. Now

$$F_{X,Y}(x, y) = P(X \le x, Y \le y)$$

$$= \begin{cases} 0 & x < 0 \quad \text{or} \quad y < -1 \\ 0 & 0 \le x < 1 \quad \text{and} \quad -1 \le y < 1 \\ 1/2 & 0 \le x < 1 \quad \text{and} \quad y \ge 1 \\ 1/2 & x \ge 1 \quad \text{and} \quad -1 \le y < 1 \\ 1 & x \ge 1 \quad \text{and} \quad y \ge 1 \end{cases}.$$

The regions investigated above are indicated in figure 7.1. Since

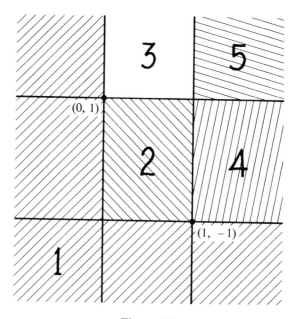

Figure 7.1

$P(X = 0, Y = 1) = 1/2 = P(X = -1, Y = -1)$ we may think of weights of $1/2$ at $(0, 1)$ and $(1, -1)$. $F_{X,Y}(x, y)$ is simply the sum of all the weights which are (at the same time) to the left of x or equal to x and below y or equal to y. We shall pursue this line further when we discuss jointly discrete random variables.

Just as all the distribution functions for random variables shared certain properties, the joint distribution functions all satisfy certain properties.

THEOREM 1:   Let $F_{X,Y}$ be the joint distribution function for the random variables X and Y on the p.m.s. (S, Σ, P). Then

(a) $F_{X,Y}: R^2 \to [0, 1]$

(b)   If $x_1 \le x_2$ and $b \in R$, then

$$F_{X,Y}(x_1, b) \le F_{X,Y}(x_2, b)$$

(c)   If $y_1 \le y_2$ and $a \in R$, then

$$F_{X,Y}(a, y_1) \le F_{X,Y}(a, y_2)$$

(d)   If $x_1 \le x_2$ and $y_1 \le y_2$, then

$$F_{X,Y}(x_2, y_2) - F_{X,Y}(x_2, y_1) - F_{X,Y}(x_1, y_2) + F_{X,Y}(x_1, y_1) \ge 0$$

(e)   For each $a \in R$,

$$\lim_{x \to -\infty} F_{X,Y}(x, a) = 0 = \lim_{y \to -\infty} F_{X,Y}(a, y)$$

(f) $\lim_{\substack{x \to \infty \\ y \to \infty}} F_{X,Y}(x, y) = 1$

(g)   If $(a, b) \in R^2$, then $\lim_{\substack{x \to a^+ \\ x \to b^+}} F_{X,Y}(x, y) = F_{X,Y}(a, b)$.

Proof:

(a)   For each $(x, y) \in R^2$, $F_{X,Y}(x, y)$ is the probability of something and this must be between 0 and 1.

(b) and (c) are simple consequences of the fact that if $A, B \in \Sigma$ and $A \subseteq B$, then $P(A) \le P(B)$.

(d) $0 \le P(x_1 < X \le x_2$ and $y_1 < Y \le y_2)$ which is the probability that $(X(s), Y(s))$ lie in the rectangle shown in figure 7.2.

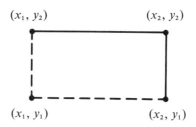

Figure 7.2

This probability can be computed by computing the probability that $(X(s), Y(s))$ lie to the left and below (x_2, y_2), then subtracting off the probability that it lie to the left and below (x_1, y_2) and that it lie to the left and below (x_2, y_1). However, the probability that it lie to the left and below (x_1, y_1) has been subtracted twice, hence it must be added back. This is illustrated in figure 7.3.

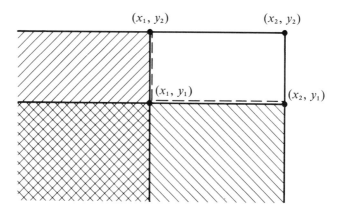

Figure 7.3

(e) For each $n \in \mathbf{N}$, let $A_n = \{s \in S \mid X(s) \le -n, Y(s) \le a\}$ then $A_1 \supseteq A_2 \supseteq A_3 \cdots \supseteq$ and $\bigcap_1^\infty A_n = \varnothing$. Thus $0 = P(\varnothing) = \lim_{n \to \infty} P(A_n) = \lim_{n \to \infty} F_{X,Y}(-n, b)$; by (b), we conclude that $\lim F_{X,Y}(x, b) = 0$. The second equality in (e) is done similarly.

(f) For each $n \in \mathbf{N}$, let $A_n = \{s \in S \mid X(s) \le n, Y(s) \le n\}$, then $A_1 \subseteq A_2 \subseteq \cdots$ and $\bigcup_1^{\infty} A_n = S$. Thus $1 = P(S) = \lim_{n \to \infty} P(A_n) = \lim_{n \to \infty} F_{X,Y}(n, n)$. Part (d) can now be used to show that

$$\lim_{\substack{x \to \infty \\ y \to \infty}} F_{X,Y}(x, y) = 1.$$

(g) Let $(a, b) \in \mathbf{R}^2$ and for each $n \in \mathbf{N}$, let

$$A_n = \{s \in S \mid X(s) \le a + 1/n, Y(s) \le b + 1/n\};$$

then $A_1 \supseteq A_2 \supseteq \cdots$ and $\bigcap_1^{\infty} A_n = \{s \in S \mid X(s) \ge a, Y(s) \le b\}$. Thus

$$F_{X,Y}(a, b) = P\left(\bigcap_1^{\infty} A_n\right) = \lim_{n \to \infty} P(A_n)$$

$$= \lim_{n \to \infty} F_{X,Y}(a + 1/n, b + 1/n).$$

Again (d) is used to show that

$$\lim_{\substack{x \to a^+ \\ y \to b^+}} F_{X,Y}(x, y) = F_{X,Y}(a, b). \quad \square$$

In a preceding chapter, it was shown that if a function from $\mathbf{R}$ into $[0, 1]$ satisfies certain properties, then it is a distribution function for some random variable on some probability measure space (Theorem 2 of section 5.2). The same type theorem can be proven in the case of a function from $\mathbf{R}^2$ into $[0, 1]$; however the proof involves probability measures on the Borel subsets of $\mathbf{R}^2$, a topic which we have not discussed. We shall state simply that the properties in Theorem 1 are sufficient for $F_{X,Y}$ to be the joint distribution function for two random variables.

If the individual distribution functions for two random variables are given, it is not possible in general to determine the joint distribution function. However, if the joint distribution function is given, it *is* possible to recover the individual distributions. The following proposition indicates the fashion in which the recovery may be made; the proof is relatively simple and is left as an exercise.

Proposition 1: Let $F_{X,Y}$ *be the joint distribution function for two random variables X and Y. Then*

(a) *for each* $x \in$ R, $\lim\limits_{y \to \infty} F_{X,Y}(x, y) = F_X(x)$.

(b) *for each* $x \in$ R, $\lim\limits_{x \to \infty} F_{X,Y}(x, y) = F_Y(y)$.

F_X and F_Y are sometimes called the *marginal distributions* corresponding to $F_{X,Y}$.

Example: In the example at the first of this section, we derived a joint distribution function.

$$F_{X,Y}(x, y) = \begin{cases} 0 & \text{if } x < 0 \quad \text{or} \quad y < -1 \\ 0 & \text{if } 0 \le x < 1 \quad \text{and} \quad -1 \le y < 1 \\ 1/2 & \text{if } 0 \le x < 1 \quad \text{and} \quad y \ge 1 \\ 1/2 & \text{if } x \ge 1 \quad \text{and} \quad -1 \le y < 1 \\ 1 & \text{if } x \ge 1 \quad \text{and} \quad y \ge 1 \end{cases}$$

Thus, we find that

$$F_X(x) = \lim_{y \to \infty} F_{X,Y}(x, y) = \begin{cases} 0 & \text{if } x < 0 \\ 1/2 & \text{if } 0 \le x < 1 \\ 1 & \text{if } x \ge 1 \end{cases}$$

and

$$F_Y(y) = \lim_{x \to \infty} F_{X,Y}(x, y) = \begin{cases} 0 & \text{if } y < -1 \\ 1/2 & \text{if } -1 \le y < 1 \\ 1 & \text{if } y \ge 1 \end{cases}.$$

Example: It can easily be verified that the function F, below, is a joint distribution function.

$$F(x, y) = \begin{cases} 0 & \text{if } x < 0 \quad \text{or} \quad y < -1 \\ 1/4 & \text{if } 0 \le x < 1 \quad \text{and} \quad -1 \le y < 1 \\ 1/2 & \text{if } 0 \le x < 1 \quad \text{and} \quad y \ge 1 \\ 1/2 & \text{if } x \ge 1 \quad \text{and} \quad -1 \le y < 1 \\ 1 & \text{if } x \ge 1 \quad \text{and} \quad y \ge 1 \end{cases}.$$

We find that this joint distribution function has exactly the same marginal distributions as does the joint distribution function in the preceding example.

Exercises:

1. Verify that properties (b) and (c) hold in Theorem 1.

2. Prove Proposition 1.

3. Find the marginal distributions corresponding to the joint distributions given below

   (a) $F(x, y) = \begin{cases} 1 - e^{-x} - e^{-y} + e^{-(x+y)} & \text{if } x \geq 0 \text{ and } y \geq 0 \\ 0 & \text{if not} \end{cases}$

   (b) $F(x, y) = \begin{cases} 0 & \text{if } x \leq 0 \quad \text{or} \quad y \leq 0 \\ x^2 y & \text{if } 0 \leq x \leq 1 \quad \text{and} \quad 0 \leq y \leq 1 \\ x^2 & \text{if } 0 \leq x \leq 1 \quad \text{and} \quad y \geq 1 \\ y & \text{if } x \geq 1 \quad \text{and} \quad 0 \leq y \leq 1 \\ 1 & \text{if } x \geq 1 \quad \text{and} \quad y \geq 1 \end{cases}$

4. Concoct a reasonable definition for the joint distribution function for n random variables $X_1, X_2, \ldots, X_n$ on a probability measure space (S, Σ, P).

5. State and prove a theorem similar to Theorem 1 for the joint distribution function for n random variables (good luck with part (d)!).

6. State and prove a proposition similar to Proposition 1 for the joint distribution function for n random variables.

7.2 Classification of Jointly Distributed Random Variables

In the case of a single random variable we saw that the distribution function often could be given in terms of a mass function or a density. We shall consider similar cases for jointly distributed random variables.

For a discrete random variable X, we considered

$$P(X = x) = F_X(x) - F_X(x^-);$$

for jointly distributed discrete random variables, we shall be interested in $P(X = x \text{ and } Y = y)$. The formula for $P(X = x, Y = y)$ in terms of the joint distribution function is given in the proposition which follows.

Proposition 1: *Let X and Y be jointly distributed random variables; then for each $(a, b) \in \mathbb{R}^2$,*

$$P(X = a, Y = b) = F_{X,Y}(a, b) - F_{X,Y}(a, b^-) - F_{X,Y}(a^-, b) + F_{X,Y}(a^-, b^-)$$

(where $F_{X,Y}(a, b^-) = \lim_{y \to b^-} F_{X,Y}(a, y)$, etc.).

Proof: For each $n \in \mathbb{N}$, let

$$A_n = \{s \in S \mid a - 1/n < X(s) \leq a \text{ and } b - 1/n < Y(s) \leq b\}.$$

Then $A_1 \supseteq A_2 \supseteq A_3 \supseteq \cdots$ and

$$\bigcap_1^\infty A_n = \{s \in S \mid X(s) = a, Y(s) = b\}.$$

Thus

$$P(X = a, Y = b) = P\left(\bigcap_1^\infty A_n\right) = \lim_{n \to \infty} P(A_n).$$

Now by the same arguments used to establish part (d) of Theorem 1 in the preceding section, we find that

$$P(A_n) = F_{X,Y}(a, b) - F_{X,Y}(a, b - 1/n)$$
$$- F_{X,Y}(a - 1/n, b) + F_{X,Y}(a - 1/n, b - 1/n).$$

We now use parts (b), (c), and (d) of that theorem to ensure that

$$\lim_{n \to \infty} P(A_n) = F_{X,Y}(a, b) - F_{X,Y}(a, b^-)$$
$$- F_{X,Y}(a^-, b) + F_{X,Y}(a^-, b^-). \quad \square$$

For a single random variable X, we found that there exist at most a countable number of real numbers x such that $P(X = x) > 0$ and that $\sum_{x \in \mathbb{R}} P(X = x) \leq 1$. By similar methods, we can prove that for jointly distributed random variables X and Y, there exist at most a countable number of pairs of real numbers (x, y) such that $P(X = x, Y = y) > 0$ and that $\sum_{\substack{x \in \mathbb{R} \\ y \in \mathbb{R}}} P(X = x, Y = y) \leq 1$.

Definition 1:   Let $F_{X,Y}$ be the joint distribution function for random variables X and Y and for each $(x, y) \in \mathbb{R}^2$, let

$$p_{X,Y}(x, y) = P(X = x, Y = y).$$

If $F_{X,Y}$ has the property that $F_{X,Y}(a, b) = \sum_{\substack{x \leq a \\ x \leq b}} p_{X,Y}(x, y)$, then X and Y are said to be *jointly discrete* and $p_{X,Y}$ is called a *joint probability mass function*.

Note that any function $p: \mathbb{R}^2 \rightarrow [0, \infty)$ such that $p(x, y) > 0$ for only a countable number of points and $\sum_{(x,y) \in \mathbb{R}^2} p(x, y) = 1$ defines a joint distribution function by $F(a, b) = \sum_{\substack{x \leq a \\ y \leq b}} p(x, y)$.

Example: In the first example of the preceding section,

$$p_{X,Y}(x, y) = \begin{cases} 1/2 & \text{if } (x, y) = (0, 1) \text{ or } (1, -1) \\ 0 & \text{otherwise} \end{cases}.$$

Example: In the last example of the preceding section we investigated

$$F(x, y) = \begin{cases} 0 & \text{if } x < 0 \quad \text{or} \quad y < -1 \\ 1/4 & \text{if } 0 \leq x < 1 \quad \text{and} \quad -1 \leq y < 1 \\ 1/2 & \text{if } 0 \leq x < 1 \quad \text{and} \quad y \geq 1 \\ 1/2 & \text{if } x \geq 1 \quad \text{and} \quad -1 \leq y < 1 \\ 1 & \text{if } x \geq 1 \quad \text{and} \quad y \geq 1 \end{cases}$$

Now $P(X = x, Y = y) = F(x, y) - F(x^-, y) - F(x, y^-) + F(x^-, y^-)$. If F is continuous in the first variable at (x, y), then $F(x, y) = F(x^-, y)$ and $F(x, y^-) = F(x^-, y^-)$, so $P(X = x, Y = y) = 0$; similarly, if F is continuous in the second variable at (x, y), then $P(X = x, Y = y) = 0$. Thus, to find points at which

$$P(X = x, Y = y) > 0,$$

we need consider only points at which F is discontinuous in both variables. In the present example, these are

$$(0, -1), (0, 1), (1, -1), \text{ and } (1, 1).$$

We compute

$$P(X = x, Y = y) = F(x, y) - F(x^-, y) - F(x, y^-) + F(x^-, y^-)$$

$$= \begin{cases} 1/4 - 0 - 0 + 0 = 1/4 & \text{at } (0, -1) \\ 1/2 - 0 - 1/4 + 0 = 1/4 & \text{at } (0, 1) \\ 1/2 - 1/4 - 0 + 0 = 1/4 & \text{at } (1, -1) \\ 1 - 1/2 - 1/2 + 1/4 = 1/4 & \text{at } (1, 1) \\ 0 & \text{otherwise} \end{cases}.$$

Example: Let $p: R^2 \to R$ by

$$p(x, y) = \begin{cases} 1/6 & \text{at } (1, 1) \\ 1/2 & \text{at } (1, 0) \\ 1/3 & \text{at } (0, 1) \\ 0 & \text{otherwise} \end{cases} .$$

Then p determines the joint distribution function

$$F(x, y) = \begin{cases} 0 & \text{if } x < 0 \quad \text{or} \quad y < 0 \\ 0 & \text{if } 0 \le x < 1 \quad \text{and} \quad 0 \le y < 1 \\ 1/3 & \text{if } 0 \le x < 1 \quad \text{and} \quad y \ge 1 \\ 1/2 & \text{if } x \ge 1 \quad \text{and} \quad 0 \le y < 1 \\ 1 & \text{if } x \ge 1 \quad \text{and} \quad y \ge 1 \end{cases} .$$

In the case of the distribution function of a single random variable X, we saw that a probability measure P^* was generated on $\mathscr{B}(R)$ by $P^*(B) = P(X^{-1}(B))$ (see exercise 3 of section 5.2). For the joint distribution function for two random variables X and Y on the p.m.s. (S, Σ, P), a probability measure P^* is determined on $\mathscr{B}(R^2)$ by

$$P^*(B) = P((X, Y) \in B) = P(\{s \in S \mid (X(s), Y(s)) \in B).$$

Let $Z: S \to R^2$ by $Z(s) = (X(s), Y(s))$, then Z may be considered a random vector (or vector-valued random variable). Then $P^*(B) = P(Z^{-1}(B))$; we proved in section 5.4 that for each $B \in \mathscr{B}(R^2)$, we have $Z^{-1}(B) \in \Sigma$, so $P(Z^{-1}(B))$ *is* defined.

In the case of jointly discrete random variables, it may be easy to actually compute $P^*(B)$:

$$P(Z^{-1}(B)) = P((X, Y) \in B) = \sum_{(x,y) \in B} p_{X,Y}(x, y).$$

Example: Let

$$p_{X,Y}(x, y) = \begin{cases} 1/8 & \text{at } (0, 0) \\ 1\,4 & \text{at } (1, 0) \text{ and } (0, 1/2) \\ 3/8 & \text{at } (2, 2) \\ 0 & \text{otherwise} \end{cases}$$

and let us compute $P(X + Y^2 \le 1)$.

If $B = \{(x, y) \in R^2 \mid x + y^2 \le 1\}$, then B is a Borel set of R and we are asked to compute $P((X, Y) \in B)$. This is easily seen to be $1/8 + 1/4 + 1/4 = 5/8$.

Proposition 1: Let X and Y be jointly discrete random variables. Then X and Y are discrete and

(a) For each $x \in R$, $p_X(x) = \sum_{y \in R} p_{X,Y}(x, y)$

(b) For each $y \in R$, $p_Y(y) = \sum_{x \in R} p_{X,Y}(x, y)$

Proof: For each $x \in R$, let $B_x = \{(x, y) \mid y \in R\} = \{x\} \times R$. Then $B_x \in \mathscr{B}(R^2)$ and

$$P(X = x) = P((X, Y) \in B_x) = \sum_{(s,t) \in B_x} p_{X,Y}(s, t) = \sum_{t \in R} p_{X,Y}(x, t)$$

$$= \sum_{y \in R} p_{X,Y}(x, y). \quad \square$$

Example: In the preceding example, we find that

$$p_X(x) = \begin{cases} 3/8 & \text{if } x = 0 \\ 1/4 & \text{if } x = 1 \\ 3/8 & \text{if } x = 2 \\ 0 & \text{otherwise} \end{cases} \qquad p_Y(y) = \begin{cases} 3/8 & \text{if } y = 0 \\ 1/4 & \text{if } y = 1/2 \\ 3/8 & \text{if } y = 2 \\ 0 & \text{otherwise} \end{cases}.$$

An easy way to display a joint probability mass function which has only a small number of non-zero values is by a matrix as in figure 7.4:

y \ x	0	1	2
0	1/8	1/4	0
1/2	1/4	0	0
2	0	0	3/8

Figure 7.4

The individual probability mass functions may be obtained by summing across or down.

Example: A fair die is tossed ten times. Investigate the number of 1's and 3's.

For the ten tosses, let X be the number of 1's and Y be the number of 3's. Then for $n = 0, 1, \ldots, 10$ and $m = 0, 1, \ldots, 10 - n$, we have that

$$P(X = n, Y = m) = \frac{10!}{n! \, m! \, (10 - n - m)!} \left(\frac{1}{6}\right)^n \left(\frac{1}{6}\right)^m \left(\frac{2}{3}\right)^{10 - n - m}.$$

We wish now to consider a concept for jointly absolutely continuous random variables. For a single random variable, we required that the density be a piecewise continuous function on R. For two random variables, we would like to have a similar requirement but we have a problem deciding what the pieces should be. For this, we shall require a few definitions. The reader who is uninterested in the mathematical details may skip to Definition 5.

Definition 2:   Let $A \subseteq R^2$; we shall say that A has (Jordan) content 0 provided that for each $\varepsilon > 0$, there is a finite set of rectangles

$$\{R_1, R_2, \ldots, R_n\} \text{ in } R^2$$

such that

(a) $A \subseteq R_1 \cup R_2 \cup \cdots \cup R_n$ and
(b) the sum of the areas of the n rectangles is less than ε.

It is easy to see that any set of content 0 must be bounded.

Example: A bounded line segment has content 0.

We erect rectangles as shown in figure 7.5.

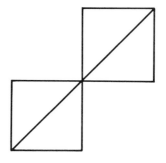

Figure 7.5

Then we split each rectangle into four congruent rectangles, as in figure 7.6.

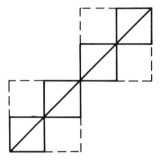

Figure 7.6

Now only two of the smaller rectangles in each large rectangle actually intersect the line segment, so the other two may be discarded. The sum of the areas of the new rectangles is half the total area of the former rectangles. This process may be repeated indefinitely. If $\varepsilon > 0$, then eventually a set of rectangles will be produced which cover A and have total area less than ε.

Essentially, a set of content 0 is one which has no "area." It is not hard to show that if A and B have content 0, then so does $A \cup B$ (use an $\varepsilon/2$ argument). Through an argument similar to that given in the preceding example, one may show that if $f: [a, b] \to$ R such that f is continuous, then the graph of f ($\{(x, f(x)) \mid x \in [a, b]\}$) has content 0. (Caution: if you attempt this, you may need the concept of uniform continuity.)

Example: The rim of a circle has content 0.

Let the circle be given, for simplicity, by $\{(x, y) \mid x^2 + y^2 \le r^2\}$; then the rim is $\{(x, y) \mid x^2 + y^2 = r^2\}$. Define $f: [-r, r] \to$ R by $f(x) = \sqrt{r^2 - x^2}$; then the graph of f, say A, is the upper half of the rim and hence this half has content 0. Define $g: [-r, r] \to$ R by $g(x) = -\sqrt{r^2 - x^2}$; then the graph of g, say B, is the lower half of the rim and hence this half has content 0. The rim is $A \cup B$.

Definition 3: Let $A \subseteq$ R^2, then the boundary of A, denoted by $\partial(A)$, is the set of all points $(x, y) \in$ R^2 such that each open rectangle which contains (x, y) also contains points of both A and R$^2 \backslash A$.

Example: Let $A = (0, 1] \times (0, 1]$.

The boundary of A consists of the four line segments indicated in figure 7.7.

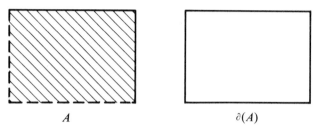

A $\partial(A)$

Figure 7.7

The main reason for introducing the concepts of content 0 and boundary is the following theorem whose proof may be found on page 324 of *The Elements of Real Analysis* by Robert G. Bartle.

THEOREM 1:   Let $A \subseteq R^2$ be bounded and $\partial(A)$ have content 0. Let $f: A \cup \partial(A) \to R$ be continuous, then f is (Riemann) integrable over A.

Example: Let $A = (0, 1] \times (0, 1]$ and $f(x, y) = x + y$.

We saw in the preceding example that $\partial(A)$ consists of four bounded line segments; since each of these line segments has content 0 then so does $\partial(A)$. The function f is continuous $A \cup \partial(A)$, so f is integrable over A. In this case, we may compute the integral as iterated integrals:

$$\iint_A f = \int_0^1 \left(\int_0^1 f(x, y) \, dy \right) dx = \int_0^1 \left(\int_0^1 x + y \, dy \right) dx$$

$$= \int_0^1 x + \frac{1}{2} \, dx = \frac{1}{2} + \frac{1}{2} = 1.$$

We now introduce a concept of improper integral for functions defined on subsets of R^2. A set $A \subseteq R^2$ will be called *admissible* provided that whenever B is any bounded subset of R^2, then $\partial(A) \cap B$ has content 0. In particular, if A is bounded, then A is admissible if and only if $\partial(A)$ has content 0.

Example: Let $A = (0, 1] \times R$ (see figure 7.8).

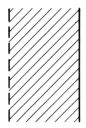

Figure 7.8

The boundary of A consists of two unbounded line segments: $\{0\} \times$ R and $\{1\} \times$ R. Now $\partial(A)$ does not have content 0, since it is unbounded; however, if B is any bounded subset of R^2, then $\partial(A) \cap B$ is a subset of two bounded line segments and hence has content 0. Thus A is admissible.

Let A be an admissible subset of R^2 and $f: A \cup \partial(A) \to$ R; then we shall say that f is integrable over A if there exists a number L such that whenever $\{A_n\}_1^\infty$ is a non-decreasing sequence of bounded admissible subsets of A such that $\bigcup_1^\infty A_n = A$, then $\lim_{n \to \infty} \iint_{An} f = L$. In this case, we shall write $\iint_A f = L$. It can be shown in particular that if A is admissible and f is non-negative and continuous on A, then $\iint_A f$ exists provided that $\{\iint_B f \mid B$ is bounded admissible subset of $A\}$ is bounded. In this case $\iint_A f$ usually may be written as iterated integrals.

The admissible subsets of R^2 will form our "pieces" for piecewise continuity.

Definition 4: A function $f: R^2 \to$ R is said to be *piecewise continuous* provided that there exists a disjoint sequence $\{A_n\}_1^\infty$ of subsets of R^2 such that

(a) $\bigcup_1^\infty A_n = R^2$,

(b) each A_n is admissible,

(c) f is continuous on each A_n.

We are now ready to proceed with our consideration of jointly absolutely continuous random variables.

Definition 5: Let $F_{X,Y}$ be the joint distribution function for random variables X and Y. X and Y are said to be *jointly absolutely continuous* provided that there exists a non-negative, piecewise continuous function $f_{X,Y}$ from $\mathbb{R}^2$ into $\mathbb{R}$ such that for each $(a, b) \in \mathbb{R}^2$,

$$F_{X,Y}(a, b) = \int_{-\infty}^{a} \int_{-\infty}^{b} f_{X,Y}(x, y)\, dy\, dx.$$

$f_{X,Y}$ is called the *joint density* for X and Y.

The joint density is determined to a large degree by the distribution function as is seen in the following proposition.

Proposition 2:   Let $F_{X,Y}$ be the joint distribution function for two jointly absolutely continuous random variables. If $f_{X,Y}$ is continuous in an open rectangle containing (a, b), then

$$f_{X,Y}(a, b) = \frac{\partial^2 F_{X,Y}(a, b)}{\partial y\, \partial x}.$$

Proof: Let R be the open rectangle containing (a, b) and let $(c, d) \in R$. Then for all $(s, t) \in R$,

$$\int_{c}^{s} \int_{d}^{t} f_{X,Y}(x, y)\, dy\, dx = F_{X,Y}(s, t) - F_{X,Y}(c, t)$$
$$- F_{X,Y}(s, d) + F_{X,Y}(c, d)$$

(to show this, look at possible relationships between s and c and between t and d).

$$\frac{\partial}{\partial x} \left(\int_{c}^{s} \int_{d}^{t} f_{X,Y}(x, y)\, dy\, dx \right) = \frac{\partial}{\partial x} F_{X,Y}(s, t) - 0$$
$$- \int_{-\infty}^{d} f_{X,Y}(s, y)\, dt + 0$$

But $\dfrac{\partial}{\partial x} \left(\int_{c}^{s} \int_{d}^{t} f_{x,y}(x, y)\, dy\, dx \right) = \int_{d}^{t} f_{X,Y}(s, y)\, dt$ also.

$$\frac{\partial}{\partial y} \left(\int_{d}^{t} f_{X,Y}(s, y)\, dy \right) = \frac{\partial}{\partial y} \left(\frac{\partial}{\partial x} F_{X,Y}(s, t) \right) - 0$$

$$|| \qquad\qquad\qquad\qquad ||$$

$$f_{X,Y}(s, t) \qquad\qquad \frac{\partial^2}{\partial y\, \partial x} F_{X,Y}(s, t).$$

In particular, when $(s, t) = (a, b)$, the desired result is obtained.   ☐

If $A = (-\infty, a] \times (-\infty, b]$ and X and Y are jointly absolutely continuous, then

$$P((X, Y) \in A) = \int\int_A f_{X,Y}(x, y)\, dy\, dx.$$

It can be shown, from the fact that two probability measures that agree on sets of the form $(-\infty, a] \times (-\infty, b]$ are the same, that if A is a Borel set in R^2 and A is admissible, then $P((X, Y) \in A) = \int\int_A f_{X,Y}(x, y)\, dy\, dx$. From this fact, we derive the following proposition.

Proposition 3: *If X and Y are jointly absolutely continuous, then X and Y are absolutely continuous and*

(1) $\int_{-\infty}^{\infty} f_{X,Y}(x, y)\, dy = f_X(x)$ *is a density for X (called a marginal density).*

(2) $\int_{-\infty}^{\infty} f_{X,Y}(x, y)\, dx = f_Y(y)$ *is a density for Y.*

Proof: We shall do only (1) since (2) is similar.

$$F_X(x) = P(X \le x) = P((X, Y) \in A) \quad \text{where} \quad A = \{(s, t) \in \mathsf{R}^2 \mid s \le x\}.$$

Then

$$P((X, Y) \in A) = \int\int_A f_{X,Y}(s, t)\, dt\, ds$$

$$= \int_{-\infty}^{x} \left(\int_{\infty}^{\infty} f_{X,Y}(s, t)\, dt \right) ds,$$

so $\int_{-\infty}^{\infty} f_{X,Y}(s, t)\, dt$ is a density for F_X. $\square$

It may be proven easily that the joint distribution function for jointly absolutely continuous random variables is continuous. However, as in the case of a single r.v., $F_{X,Y}$ may be continuous but X and Y not be absolutely continuous. Even if $F_{X,Y}$ is continuous and has continuous second partials piecewise in R^2, X and Y need not be jointly absolutely continuous (see exercise 4). The author is not aware of any quick, simple test of a joint distribution function which ensures that it be the distribution function for jointly absolutely continuous random variables.

If $f: \mathbb{R}^2 \to \mathbb{R}$ is any piecewise continuous function such that $f(x, y) \geq 0$ for each $(x, y) \in \mathbb{R}^2$, and $\int_{-\infty}^{\infty} \int_{-\infty}^{\infty} f(x, y) \, dy \, dx = 1$, then f determines a joint distribution for jointly absolutely continuous random variables by

$$F(a, b) = \int_{-\infty}^{a} \int_{-\infty}^{b} f(x, y) \, dy \, dx.$$

Example: Let

$$f(x, y) = \begin{cases} 4xy & \text{on } (0, 1) \times (0, 1) \\ 0 & \text{otherwise} \end{cases}.$$

Then f is a joint density since

$$\int_{-\infty}^{\infty} \int_{-\infty}^{\infty} f(x, y) \, dy \, dx = \int_{0}^{1} \int_{0}^{1} 4xy \, dy \, dx = 1.$$

One marginal density is given by

$$f_1(x) = \int_{-\infty}^{\infty} f(x, y) \, dy = \begin{cases} \int_{0}^{1} 4xy \, dy & \text{if } 0 < x < 1 \\ 0 & \text{if not} \end{cases}$$

$$= \begin{cases} 2x & \text{if } 0 < x < 1 \\ 0 & \text{if not} \end{cases},$$

and the other density is similar.

Example: Let $f: \mathbb{R}^2 \to \mathbb{R}$ by

$$f(x, y) = \begin{cases} x + y & \text{if } 0 < x < 1 \text{ and } 0 < y < 1 \\ 0 & \text{otherwise} \end{cases}$$

f is piecewise continuous and

$$\int_{-\infty}^{\infty} \int_{-\infty}^{\infty} f(x, y) \, dy \, dx = \int_{0}^{1} \int_{0}^{1} x + y \, dy \, dx = 1$$

so f is a joint density. The joint distribution function generated by f is

$$F(x, y) = \int_{-\infty}^{x} \int_{-\infty}^{y} f(s, t) \, dt \, ds$$

$$= \begin{cases} 0 & \text{if } x < 0 \quad \text{or} \quad y < 0 \\ \int_{0}^{x} \int_{0}^{y} (s + t) \, dt \, ds & \text{if } 0 \le x \le 1 \quad \text{and} \quad 0 \le y < 1 \\ \int_{0}^{x} \int_{0}^{1} (s + t) \, dt \, ds & \text{if } 0 \le x < 1 \quad \text{and} \quad y \ge 1 \\ \int_{0}^{1} \int_{0}^{y} (s + t) \, dt \, ds & \text{if } x \ge 1 \quad \text{and} \quad 0 \le y < 1 \\ \int_{0}^{1} \int_{0}^{1} (s + t) \, dt \, ds & \text{if } x \ge 1 \quad \text{and} \quad y \ge 1 \end{cases}$$

$$= \begin{cases} 0 & \text{if } x < 0 \quad \text{or} \quad y < 0 \\ (x^2 y + y^2 x)/2 & \text{if } 0 \le x < 1 \quad \text{and} \quad 0 \le y < 1 \\ (x^2 + x)/2 & \text{if } 0 \le x < 1 \quad \text{and} \quad y \ge 1 \\ (y^2 + y)/2 & \text{if } x \ge 1 \quad \text{and} \quad 0 \le y < 1 \\ 1 & \text{if } x \ge 1 \quad \text{and} \quad y \ge 1 \end{cases}$$

One marginal density is

$$f_1(x) = \int_{-\infty}^{\infty} f(x, y) \, dy = \begin{cases} \int_{0}^{1} x + y \, dy & \text{if } 0 < x < 1 \\ 0 & \text{otherwise} \end{cases}$$

$$= \begin{cases} x + 1/2 & \text{if } 0 < x < 1 \\ 0 & \text{otherwise} \end{cases}$$

Exercises:

1. X and Y are jointly discrete with joint probability mass function

$$p_{X,Y}(x, y) = \begin{cases} 1/5 & \text{at } (0, 1) \text{ and } (2, 1) \\ 3/5 & \text{at } (1, 1) \\ 0 & \text{otherwise} \end{cases}$$

Find the individual probability mass functions.

2. Let $\lambda > 0$ and $\mu > 0$. Find C such that

$$p(x, y) = \begin{cases} C \dfrac{\lambda^x \mu^y}{x! \, y!} & \text{if } x = 0, 1, 2, 3, \ldots \\ & \text{and } y = 0, 1, 2, 3, \ldots \\ 0 & \text{otherwise} \end{cases}$$

is a joint probability mass function. Find the individual probability mass functions.

3. Each of the following is a joint density. Find C. Find F. Find the individual densities.

    (a) $f(x, y) = Ce^{-((x^2 + y^2)/2)}$

    (b) $f(x, y) = \begin{cases} C & \text{on } [a, b] \times [c, d] \\ 0 & \text{otherwise} \end{cases}$

    (c) $f(x, y) = \begin{cases} A(x^2 + y^2) & \text{on } [0, 1] \times [0, 1] \\ 0 & \text{otherwise} \end{cases}$

    (d) $f(x, y) = \begin{cases} A(1 - (x^2 + y^2)) & \text{if } x^2 + y^2 \le 1 \\ 0 & \text{otherwise} \end{cases}$.

4. Let X be an absolutely continuous random variable with density

$$f_X(x) = \begin{cases} 1 & \text{on } (0, 1) \\ 0 & \text{otherwise} \end{cases} .$$

Let $Y = X$. Show that X and Y are *not* jointly absolutely continuous!

5. A garment manufacturer makes shirts of two qualities, firsts and seconds, as well as some rejects. The qualities appear to occur at random with 75 percent firsts, 15 percent seconds, and 10 percent rejects. A sample of eight shirts is drawn from the output of the manufacturer. Let X be the number of firsts and Y be the number of seconds. Find $p_{X,Y}$.

6. An archer who considers himself fairly accurate shoots at a target one foot in radius. Treat the point of contact for the arrow as a random vector (X, Y) which has a joint density given as in problem 3(d) above. Find the probability that the arrow is closer to the center than to the edge.

7.3 Independent Random Variables

Let (S, Σ, P) be a p.m.s. and $X: S \rightarrow \mathsf{R}$. We have seen that sets of the form $\{s \in S \mid X(s) \leq x\}$ are very important; they determine whether X is a random variable, and if so, then they determine the distribution function and in case of two or more random variables, the joint distribution. The concept of independence of random variables also will be given in terms of these sets.

Definition 1: Let (S, Σ, P) be a probability measure space and X and Y be random variables on S. X and Y are *independent* provided that for each $x \in \mathsf{R}$ and each $y \in \mathsf{R}$, $\{s \in S \mid X(s) \leq x\}$ and $\{s \in S \mid Y(s) \leq y\}$ are independent.

In terms of probability, the definition of independence of X and Y requires that for each $x \in \mathsf{R}$ and $y \in \mathsf{R}$,

$$P(X \leq x \text{ and } Y \leq y) = P(X \leq x) \cdot P(Y \leq y).$$

When we translate the terms in this equation into distribution functions, we obtain

$$F_{X,Y}(x, y) = F_X(x) \cdot F_Y(y).$$

For discrete and absolutely continuous random variables we also can derive a criterion for independence in terms of the mass function or density.

THEOREM 1: Let X and Y be jointly discrete random variables. These are equivalent:

(a) X and Y are independent

(b) For each $(x, y) \in \mathsf{R}^2$, $p_{X,Y}(x, y) = p_X(x) \cdot p_Y(y)$.

Proof:

(a) $\Rightarrow$ (b) If X and Y are independent, then

$$
\begin{aligned}
p_{X,Y}(x, y) &= F_{X,Y}(x, y) - F_{X,Y}(x, y^-) - F_{X,Y}(x^-, y) \\
&\quad + F_{X,Y}(x^-, y^-) \\
&= F_X(x)F_Y(y) - F_X(x)F_Y(y^-) \\
&\quad - F_X(x^-)F_Y(y) - F_X(x^-)F_Y(y^-) \\
&= F_X(x)[F_Y(y) - F_Y(y^-)] - F_X(x^-)[F_Y(y) - F_Y(y^-)] \\
&= [F_X(x) - F_X(x^-)] \cdot [F_Y(y) - F_Y(y^-)] \\
&= p_X(x) \cdot p_Y(y).
\end{aligned}
$$

(b) $\Rightarrow$ (a)   If $p_{X,Y}(x, y) = p_X(x) \cdot p_Y(y)$ for each $(x, y) \in R^2$, then

$$F_{X,Y}(a, b) = \sum_{\substack{x \le a \\ y \le b}} p_{X,Y}(x, y) = \sum_{\substack{x \le a \\ y \le b}} p_X(x) p_Y(y)$$

$$= \sum_{x \le a} \left(\sum_{y \le b} p_X(x) p_Y(y) \right)$$

$$= \left(\sum_{x \le a} p_X(x) \right) \left(\sum_{y \le b} p_Y(y) \right) = F_X(a) \cdot F_Y(b). \quad \square$$

For jointly absolutely continuous random variables, we get almost the same result regarding densities—problems arise only because a density is not uniquely determined everywhere.

THEOREM 2:   Let X and Y be jointly absolutely continuous random variables.

(a)   Let X and Y be independent and $(x, y) \in R^2$; if $f_{X,Y}$ is continuous at (x, y), if f_X is continuous at x and f_Y is continuous at y, then

$$f_{X,Y}(x, y) = f_X(x) \cdot f_Y(y).$$

(b)   If $f_{X,Y}(x, y) = f_X(x) \cdot f_Y(y)$ for (x, y) in R^2, then X and Y are independent.

Proof:

(a)   If X and Y are independent, then at points of continuity of the densities,

$$f_{X,Y}(x, y) = \frac{\partial^2}{\partial y \, \partial x} (F_{X,Y}(x, y)) = \frac{\partial^2}{\partial y \, \partial x} (F_X(x) F_Y(y))$$

$$= \frac{\partial}{\partial y} (f_X(x) F_Y(y)) = f_X(x) f_Y(y).$$

(b)   For each $(a, b) \in R^2$,

$$F_{X,Y}(a, b) = \int_{-\infty}^{a} \int_{-\infty}^{b} f_{X,Y}(x, y)\, dy\, dx$$

$$= \int_{-\infty}^{a} \int_{-\infty}^{b} f_X(x)f_Y(y)\, dy\, dx$$

$$= \int_{-\infty}^{a} f_X(x) \left(\int_{-\infty}^{b} f_Y(y)\, dy \right) dx$$

$$= \int_{-\infty}^{a} f_X(x)\, (F_Y(b))\, dx$$

$$= \left(\int_{-\infty}^{a} f_X(x)\, dx \right) (F_Y(b))$$

$$= F_X(a)F_Y(b). \quad \square$$

Example: X and Y are independent discrete random variables with probability mass functions

$$p_X(x) = \begin{cases} (1/2)^x & \text{if } x = 1, 2, 3, \ldots \\ 0 & \text{otherwise} \end{cases}$$

$$p_Y(y) = \begin{cases} (1/6)(5/6)^{y-1} & \text{if } y = 1, 2, 3, \ldots \\ 0 & \text{if not} \end{cases}.$$

Since X and Y are independent, then $p_{X,Y}(x, y) = p_X(x)p_Y(y)$ so we compute

$$p_{X,Y}(x, y) = \begin{cases} (1/2)^x(1/6)(5/6)^{y-1} & \text{if } x = 1, 2, 3, \ldots \\ & \text{and } y = 1, 2, 3, \ldots . \\ 0 & \text{if not} \end{cases}$$

We may compute, for example,

$$P(X + Y = 3) = \sum_{x+y=3} p_{X,Y}(x, y)$$

$$= P_{X,Y}(1, 2) + P_{X,Y}(2, 1)$$

$$= (1/2)^1(1/6)(5/6)^1 + (1/2)^2(1/6)(5/6)^0$$

$$= 1/9.$$

Example: Two people agree to be at a certain point between 3:00 and 4:00 one day and each agrees to wait for the other for fifteen minutes. What is the probability that they will meet? How long should they agree to wait so that the probability of their meeting be at least 0.9?

Let X be the random variable describing the time at which the first person arrives and Y, the time the second person arrives. We shall assume that X and Y are independent and have uniform densities

$$f_X(x) = f_Y(y) = \begin{cases} 1 & \text{on } [3, 4] \\ 0 & \text{otherwise} \end{cases}.$$

Then the joint density is

$$f_{X,Y}(x, y) = f_X(x)f_Y(y) = \begin{cases} 1 & \text{on } [3, 4] \times [3, 4] \\ 0 & \text{otherwise} \end{cases}.$$

We seek $P(|X - Y| \le 1/4)$ since fifteen minutes $= 1/4$ hour. We shall actually compute $P(|X - Y| \le a)$ since this will help us to answer the second part. The rectangle $[3, 4] \times [3, 4]$ is illustrated in figure 7.9; also illustrated is the region $A = \{(x, y) \mid |x - y| \le a\}$ for $0 \le a \le 1$. Since the surface $z = f_{X,Y}(x, y)$ is flat at a height 1 above A, then $P((X, Y) \in A) = \iint_A f_{X,Y} = $ area A. The area (of the square) not in A is actually easier to compute: $2((1/2)(1 - a)^2) = (1 - a)^2$ so $P(|X - Y| \le a) = (1 - (1 - a)^2)$. For $a = 1/4$, we find that the probability of their meeting is $1 - (3/4)^2 = 1 - 9/16 = 7/16$. For the second question, we must solve the inequality

$$0.9 \le 1 - (1 - a)^2;$$

we get $a \ge 1 - \sqrt{10}/10 = 1 - 0.3162 = 0.6838$, which is about forty-one minutes.

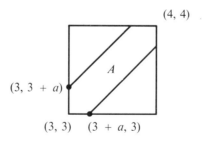

Figure 7.9

If X and Y are independent random variables and $A = (-\infty, a]$ and $B = (-\infty, b]$, then

$$P((X, Y) \in A \times B) = P(X \in A) \cdot P(Y \in B).$$

It is also true (and surprisingly hard to prove) that if A and B are Borel sets of R, then

$$P((X, Y) \in A \times B) = P(X \in A) \cdot P(Y \in B).$$

The proof involves showing that a probability measure on $\mathscr{B}(\mathsf{R}^2)$ is determined by the value of the measure on sets of the form $(-\infty, a] \times (-\infty, b]$ (see Paul R. Halmos' *Measure Theory*, 1950, pp. 54 ff., published by Van Nostrand).

We have already seen that the composition of a random variable and a Borel function is again a random variable. It is interesting, and often useful, that such compositions preserve independence.

THEOREM 3:   Let X and Y be independent random variables on a probability measure space (S, Σ, P) and let f and g be Borel functions from R into R. Then $f \circ X$ and $g \circ Y$ are independent.

Proof: Let $Z = f \circ X$ and $W = g \circ Y$; let $a \in \mathsf{R}$ and $b \in \mathsf{R}$; we must show that

$$P(Z \leq a \text{ and } W \leq b) = P(Z \leq a) \cdot P(W \leq b).$$

Let $A = (-\infty, a]$ and $B = (-\infty, b]$, then we recall that since f and g are Borel functions, then $f^{-1}(A) \in \mathscr{B}(\mathsf{R})$ and $g^{-1}(B) \in \mathscr{B}(\mathsf{R})$.   Now

$$\begin{aligned} \{s \in S \mid Z(s) \leq a, W(s) \leq b\} &= \{s \in S \mid Z(s) \in A, W(s) \in B\} \\ &= \{s \in S \mid f(X(s)) \in A, g(Y(s)) \in B\} \\ &= \{s \in S \mid X(s) \in f^{-1}(A), Y(s) \in g^{-1}(B)\}. \end{aligned}$$

Thus

$$\begin{aligned} (P(Z \leq a, W \leq b) &= P(X \in f^{-1}(A), Y \in g^{-1}(B)) \\ &= P((X, Y) \in f^{-1}(A) \times g^{-1}(B)) \\ &= P(X \in f^{-1}(A)) \cdot P(Y \in g^{-1}(B)) \\ &= P(f \circ X \in A) \cdot P(g \circ Y \in B) \\ &= P(Z \leq a) \cdot P(Y \leq b). \quad \square \end{aligned}$$

Example: Let X and Y be independent absolutely continuous random variables with densities

$$f_X(x) = f_Y(y) = \begin{cases} 1 & \text{on } [0, 1] \\ 0 & \text{otherwise} \end{cases}.$$

Let $Z = X^2$ and $W = Y^3$.

By the preceding result, Z and W are independent. We shall find the joint density for Z and W. By Theorem 2, $f_{Z,W}(z, w) = f_Z(z) \cdot f_W(w)$ so we need only find $f_Z(z)$ and $f_W(w)$.

$$f_Z(z) = \frac{d}{dz}(F_Z(z)) = \frac{d}{dz}(P(Z \le z))$$

$$= \frac{d}{dz}(P(X^2 \le z))$$

$$= \begin{cases} \dfrac{d}{dz}(P(X \le \sqrt{z})) & \text{for } z > 0 \\ 0 & \text{for } z \le 0 \end{cases}$$

$$= \begin{cases} \dfrac{d}{dz} F_X(\sqrt{z}) & \text{for } z > 0 \\ 0 & \text{for } z \le 0 \end{cases}$$

$$= \begin{cases} f_X(\sqrt{z})1/2(1/\sqrt{z}) & \text{for } z > 0 \\ 0 & \text{for } z \le 0 \end{cases}$$

$$= \begin{cases} 1/(2\sqrt{z}) & \text{for } 0 < z < 1 \\ 0 & \text{otherwise} \end{cases}.$$

$$f_W(w) = \frac{d}{dw}(F_W(w)) = \frac{d}{dw}(P(W \le w))$$

$$= \frac{d}{dw}(P(Y^3 \le w))$$

$$= \frac{d}{dw}(P(Y \le \sqrt[3]{w}))$$

$$= \left(\frac{d}{dw}\right) F_Y(\sqrt[3]{w})$$

$$= f_Y(\sqrt[3]{w})1/3(1/\sqrt[3]{w^2})$$

$$= \begin{cases} 1/(3\sqrt[3]{w^2}) & \text{if } 0 < w < 1 \\ 0 & \text{otherwise} \end{cases}.$$

Thus

$$f_{Z,W}(z, w) = \begin{cases} \dfrac{1}{6\sqrt{z}\,\sqrt[3]{w^2}} & \text{if } (z, w) \in (0, 1) \times (0, 1) \\ 0 & \text{otherwise} \end{cases}$$

Example: (Buffon's Needle Problem) A table has equidistant parallel lines ruled at a distance L apart. A needle of length l is dropped at random on the table (assume $l < L$). Find the probability that a line is crossed by the needle.

Let X be a r.v. giving the distance from the center of the needle to the nearest line. Let Y be a r.v. giving the acute angle that the needle makes with the lines (see figure 7.10). We shall assume that X and Y

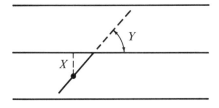

Figure 7.10

are independent and that X is uniformly distributed on $[0, L/2]$ and Y is uniformly distributed on $[0, \pi/2]$. Now the needle intersects the nearest line if and only if $X \le l/2 \sin Y$ so

$$P(\text{needle intersects the line}) = P(X \le l/2 \sin Y) = P((X, Y) \in B)$$

where B is the Borel set $\{(x, y) \mid x \le l/2 \sin y\}$. Thus

$$P\left(X \le \frac{l}{2} \sin Y\right) = \iint_B f_{X,Y} = \int_0^{\pi/2} \int_0^{l/2 \sin y} \frac{2}{l} \cdot \frac{2}{\pi}\, dx\, dy$$

$$= \frac{4}{L\pi} \int_0^{\pi/2} \frac{l}{2} \sin y\, dy$$

$$= \frac{2l}{L\pi}.$$

In this section, we have treated only two random variables at a time; actually, the results hold *mutatis mutandis* for any finite number of random variables. For example, the definition of independence for n random variables $X_1, X_2, \ldots, X_n$ would require that for any $a_1, a_2, \ldots, a_n \in \mathbb{R}$, the sets $X_1^{-1}((-\infty, a_1]), X_2^{-1}((-\infty, a_2]), \ldots, X_n^{-1}((-\infty, a_n])$ be independent.

Exercises:

1. X and Y are independent discrete random variables with probability mass functions

    $$p_X(x) = \begin{cases} e^{-\lambda}\lambda^x/x! & \text{if } x = 0, 1, 2, \ldots \\ 0 & \text{otherwise} \end{cases},$$

    $$p_Y(y) = \begin{cases} e^{-\mu}\mu^y/y! & \text{if } 0, 1, 2, \ldots \\ 0 & \text{otherwise} \end{cases}.$$

    where $\lambda > 0$ and $\mu > 0$. Find the joint probability mass function.

2. X and Y are jointly discrete with joint probability mass function:

    $$p_{X,Y}(x, y) = \begin{cases} 1/4 & \text{at } (0, 0), (0, 1), (1, 0), \text{ and } (1, 1) \\ 0 & \text{otherwise} \end{cases}$$

    Are X and Y independent?

3. X and Y are independent random variables with

    $$F_X(x) = \begin{cases} 0 & \text{if } x < 0 \\ x & \text{if } 0 \le x < 1 \\ 1 & \text{if } x \ge 1 \end{cases} \qquad F_Y(y) = \begin{cases} 0 & \text{if } x < 0 \\ 1 & \text{if } x \ge 0 \end{cases}$$

    Find $F_{X,Y}$ and attempt a sketch of the surface $z = F_{X,Y}(x, y)$.

4. X and Y are jointly absolutely continuous with joint density

    $$f_{X,Y}(x) = \begin{cases} (6/5)(x + y^2) & \text{on } (0, 1) \times (0, 1) \\ 0 & \text{otherwise} \end{cases}.$$

    Show that X and Y are not independent.

5. X and Y are independent absolutely continuous random variables with densities

    $$f_X(x) = f_Y(y) = \begin{cases} 1 & \text{on } (0, 1) \\ 0 & \text{otherwise} \end{cases}.$$

    Let $Z(s) = ln(X(s))$ and $W(s) = ln(Y(s))$. Find a joint density for Z and W.

6. Suppose (S, Σ, P) is a probability measure space and $X: S \to R$ is a random variable. Let $Y: S \to R$ be a constant function (say $Y(s) = c$ for $s \in S$). Show that Y is a random variable and that X and Y are independent.

7.4   Functions of Several Random Variables

In section 5.4 we proved that if $X_1, X_2, \ldots, X_n$ are random variables on a p.m.s. (S, Σ, P) and $f: R^n \to R$ is a Borel function, then $Y(s) = f((X_1(s), X_2(s), \ldots, X_n(s))$ is also a random variable. In some cases, the distribution function for Y may be computed easily from the joint distribution function of $X_1, \ldots, X_n$, but usually this computation is quite difficult. In this section we shall treat a special case in which the distribution function may be derived explicitly and then consider a more general formula for computing, not F_Y, but $E(Y)$. As before, we shall be concerned with only two random variables but our arguments may be extended easily to the case in which more than two are involved.

Proposition 1:   *Let X and Y be jointly discrete, and let $Z = X + Y$. Then the probability mass function for Z is*

$$p_Z(a) = \sum_{x \in R} p_{X,Y}(x, a - x).$$

Proof: Let $a \in R$ and

$$A = \{(x, y) \in R^2 \mid x + y = a\} = \{(x, a - x) \mid x \in R\}$$

then A is a Borel subset of R^2 so

$$P(Z = a) = P(X + Y = a) = P((X, Y) \in A) = \sum_{(x,y) \in A} p_{X,Y}(x, y)$$

$$= \sum_{x \in R} p_{X,Y}(x, a - x).$$

Now to show that p_Z is actually a probability mass function, we compute

$$\sum_{a \in R} p_Z(a) = \sum_{a \in R} \sum_{x \in R} p_{X,Y}(x, a - x)$$

which a rearrangement of

$$\sum_{(x,a) \in R^2} p_{X,Y}(x, a)$$

which has sum 1. $\square$

Proposition 2:   *Let X and Y be jointly absolutely continuous, and let $Z = X + Y$. Then a density for Z is $f_Z(a) = \int_{-\infty}^{\infty} f_{X,Y}(x, a - x)\, dx$.*

Proof: Let $a \in R$ and $A = \{(x, y) \mid x + y < a\}$, then A is a Borel subset of R^2 and

$$F_Z(a) = P(Z \leq a) = P(X + Y \leq a) = \iint_A f_{X,Y}$$

$$= \int_{-\infty}^{\infty} \int_{-\infty}^{a-x} f_{X,Y}(x, y) \, dy \, dx$$

$$= \int_{-\infty}^{\infty} \int_{-\infty}^{a} f_{X,Y}(x, t - x) \, dt \, dx \quad \text{where } t = x + y$$

$$= \int_{-\infty}^{a} \int_{-\infty}^{\infty} f_{X,Y}(x, t - x) \, dx \, dt$$

$$= \int_{-\infty}^{a} f_Z(t) \, dt.$$

So by definition, f_Z is a density for Z. $\square$

In case X and Y are independent, then $p_{X,Y}(x, y) = p_X(x) \cdot p_Y(y)$ for jointly discrete and $f_{X,Y}(x, y) = f_X(x) \cdot f_Y(y)$ (usually) for absolutely continuous random variables. Thus the formulae in the preceding propositions become

$$p_Z(a) = \sum_{x \in R} p_X(x) p_Y(a - x) \quad \text{and}$$

$$f_Z(a) = \int_{-\infty}^{\infty} f_X(x) f_Y(a - x) \, dx.$$

These expressions are sometimes called the convolutions of p_X and p_Y or of f_X and f_Y (at a) and are denoted by $(p_X * p_Y)(a)$ and $(f_X * f_Y)(a)$. Thus we have the following result.

Corollary: Let X and Y be independent random variables and $Z = X + Y$;

(a)   If X and Y are jointly discrete, then

$$P_Z(a) = (p_X * p_Y)(a) = \sum_{x \in R} p_X(x) p_Y(a - x).$$

(b)   If X and Y are jointly absolutely continuous, then

$$f_Z(a) = (f_X * f_Y)(a) = \int_{-\infty}^{\infty} f_X(x) f_Y(a - x) \, dx.$$

Example: Let X and Y be jointly absolutely continuous with joint density

$$f_{X,Y}(x, y) = \begin{cases} x + y & \text{if } 0 < x < 1 \\ & \quad 0 < y < 1. \\ 0 & \text{if not} \end{cases}$$

We know that for $Z = X + Y, f_Z(a) = \int_{-\infty}^{\infty} f_{X,Y}(x, a - x) \, dx$. Now $f_{X,Y}(x, a - x) > 0$ only when $0 < x < 1$ and $0 < a - x < 1$. This is equivalent to $0 < a < 2$, $0 < x < 1$, and $a - 1 < x < a$, in which case, $f_{X,Y}(x, a - x) = a$. Thus

$$\text{for } 0 < a < 1, \quad f_Z(a) = \int_0^a a \, dx = a^2$$

$$\text{for } 1 < a < 2, \quad f_Z(a) = \int_{a-1}^1 a \, dx = a(2 - a)$$

$$\text{otherwise} \quad f_Z(a) = 0.$$

This density is outlined in figure 7.11.

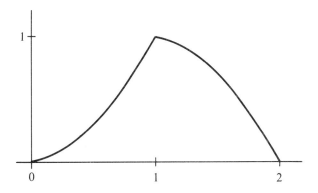

Figure 7.11

Example: Let X and Y be independent discrete random variables with Poisson probability mass functions with parameters λ and μ.

We have that

$$p_X(x) = \begin{cases} (e^{-\lambda}\lambda^x)/x! & \text{if } x = 0, 1, 2, \ldots \\ 0 & \text{if not} \end{cases}$$

$$p_Y(y) = \begin{cases} (e^{-\mu}\mu^y)/y! & \text{if } y = 0, 1, 2, \ldots \\ 0 & \text{if not} \end{cases}$$

and $p_{X,Y}(x, y) = p_X(x) \cdot p_Y(y)$. Now for

$$p_{X,Y}(x, a - x) = p_X(x)p_Y(a - x)$$

to be positive we must have $x = 0, 1, 2, \ldots$ and $a - x = 0, 1, 2, \ldots$. Thus, if $Z = X + Y$, then

$$p_Z(a) = \sum_{x \in R} p_X(x) p_Y(a - x) = \sum_{n=0}^{\infty} p_X(n) p_Y(a - n)$$

$$= \sum_{n=0}^{a} \frac{e^{-\lambda}\lambda^n}{n!} \frac{e^{-\mu}\mu^{(a-n)}}{(a - n)!}$$

$$= \frac{e^{-(\lambda+\mu)}}{a!} \sum_{n=0}^{a} \frac{a!}{n!(a - n)!} \lambda^n \mu^{(a-n)}$$

$$= \frac{e^{-(\lambda+\mu)}(\lambda + \mu)^a}{a!} \quad \text{(by the Binomial Theorem).}$$

Thus $X + Y$ has a Poisson probability mass function with parameter $\lambda + \mu$.

Example: Let X and Y be independent absolutely continuous random variables with

$$f_X(x) = f_Y(y) = \begin{cases} 1 & \text{on } (0, 1) \\ 0 & \text{otherwise} \end{cases}.$$

As before, $f_X(x)f_Y(a - x) > 0$ only when $0 < a < 2$, $0 < x < 1$, and $a - 1 < x < a$. Thus if $Z = X + Y$, then we have

$$\text{for } 0 < a < 1, \quad f_Z(a) = \int_0^a 1 \, dx = a;$$

$$\text{for } 1 < a < 2, \quad f_Z(a) = \int_{a-1}^1 1 \, dx = 2 - a;$$

$$\text{otherwise,} \quad f_Z(a) = 0.$$

The density is sketched in figure 7.12.

Example: When two resistors are connected in series, the effective resistance is the sum of the two resistances. Suppose two independent

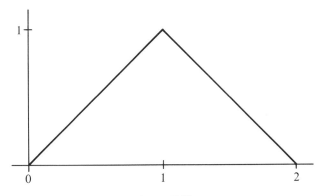

Figure 7.12

resistors have their resistances (in ohms) given by random variables X and Y such that

$$f_X(t) = f_Y(t) = \begin{cases} 1/50 & \text{if } 175 \leq t \leq 225 \\ 0 & \text{if not} \end{cases}.$$

Find the probability that a given resistor will have resistance more than 10 percent above its expectation. Find the probability that the two wired in series will have resistance more than 10 percent above the system's expectation.

$E(X) = E(Y) = 200$ and 10 percent of 200 is 20 so

$$P(X \geq 220) = \int_{220}^{\infty} f_X(t)\, dt = \int_{220}^{225} \frac{1}{50}\, dt = \frac{5}{50} = \frac{1}{10}.$$

Now to find the answer for the system of two resistors, we must investigate $Z = X + Y$.

Since X and Y are independent, then $f_Z(a) = \int_{-\infty}^{\infty} f_X(t) f_Y(a - t)\, dt$. $f_X(t) f_Y(a - t) > 0$ only when $175 \leq t \leq 225$ and $175 \leq a - t \leq 225$ which is equivalent to requiring $350 \leq a \leq 450$, $175 \leq t \leq 225$, and $a - 225 \leq t \leq a - 175$. Therefore for $350 \leq a \leq 450$,

$$f_Z(a) = \begin{cases} \displaystyle\int_{175}^{a-175} \frac{1}{50} \cdot \frac{1}{50}\, dt = \frac{1}{2500}(a - 350) & \text{for } 350 \leq a \leq 400 \\[4mm] \displaystyle\int_{a-225}^{225} \frac{1}{50} \cdot \frac{1}{50}\, dt = \frac{1}{2500}(450 - a) & \text{for } 400 \leq a \leq 450. \end{cases}$$

This density is sketched in figure 7.13 and it is easily seen that $E(Z) = 400$. Now 10 percent of 400 is 40 and

$$P(Z \geq 440) = \int_{440}^{\infty} f_Z(t)\, dt = \int_{440}^{450} \frac{1}{2500} (450 - (dt)$$

$$= 50/2500(50) = 1/50.$$

Example: Let X and Y be independent and uniformly distributed on $[0, 1]$; find a density for $Z = X \cdot Y$.

In order to use the preceding corollary, we must have a random variable expressed as the *sum* of two independent random variables. Let $Z^* = lnZ$, $X^* = lnX$, and $Y^* = lnY$ (i.e. $Z^*(s) = ln(Z(s))$, etc.); so $Z^* = X^* + Y^*$. By a result of the preceding section, X^* and Y^* are still independent. Thus

$$f_{Z*}(a) = \int_{-\infty}^{\infty} f_{X*}(x) f_{Y*}(a - x)\, dx.$$

We compute

$$f_{X*}(x) = \frac{d}{dx}(F_{X*}(x)) = \frac{d}{dx}(P(X^* \leq x))$$

$$= \frac{d}{dx}(P(lnX \leq x))$$

$$= \frac{d}{dx}(P(X \leq e^x))$$

$$= \frac{d}{dx}(F_X(e^x))$$

$$= f_X(e^x) \cdot e^x$$

$$= \begin{cases} e^x & \text{if } 0 < e^x < 1 \\ 0 & \text{otherwise} \end{cases}$$

$$= \begin{cases} e^x & \text{if } x < 0 \\ 0 & \text{otherwise} \end{cases}.$$

Similarly,

$$f_{Y*}(y) = \begin{cases} e^y & \text{if } y < 0 \\ 0 & \text{otherwise} \end{cases}.$$

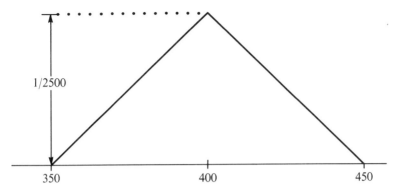

Figure 7.13

Now

$$f_{Z*}(a) = \int_{-\infty}^{\infty} f_{X*}(x) f_{Y*}(a - x) \, dx \quad \text{and} \quad f_{X*}(x) \cdot f_{Y*}(a - x) > 0$$

only when $x < 0$ and $a - x < 0$. Thus

$$f_{Z*}(a) = \begin{cases} \int_a^0 e^x e^{a-x} \, dx = \int_a^0 e^a \, dx = -ae^a & \text{if } a < 0 \\ 0 & \text{otherwise} \end{cases}$$

$$f_Z(x) = \frac{d}{dx} F_Z(x) = \frac{d}{dx} (P(Z \leq x))$$

$$= \frac{d}{dx} (P(\ln Z \leq \ln x))$$

$$= \frac{d}{dx} (P(Z^* \leq \ln x))$$

$$= \frac{d}{dx} F_{Z*}(\ln x)$$

$$= f_{Z*}(\ln x)(1/|x|)$$

$$= \begin{cases} \dfrac{-\ln x \, e^{\ln x}}{|x|} & \text{if } \ln x < 0 \\ 0 & \text{if not} \end{cases}$$

$$= \begin{cases} -\ln x & \text{if } 0 < x < 1 \\ 0 & \text{if not} \end{cases}.$$

This density is sketched in figure 7.14.

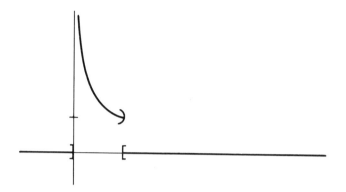

Figure 7.14

Example: Let X and Y be independent r.v.'s with normal densities $f_X(t) = f_Y(t) = (1/\sqrt{2\pi})e^{-(t^2/2)}$. Find a density for $X^2 + Y^2$.

X^2 and Y^2 are independent so we need only find f_{X^2} and f_{Y^2} in order to use Proposition 2. Let $Z = X^2$; then

$$F_X(t) = P(Z \le t) = P(X^2 \le t) = P(-\sqrt{t} \le X \le \sqrt{t})$$

$$= \begin{cases} F_X(\sqrt{t}) - F_X(-\sqrt{t}) & t \ge 0 \\ 0 & t < 0 \end{cases}.$$

Thus

$$f_Z(t) = \left(\frac{d}{dt}\right)F_Z(t) = \begin{cases} \left(\dfrac{d}{dt}\right)(F_X(\sqrt{t}) - F_X(-\sqrt{t})) & t > 0 \\ 0 & t \le 0 \end{cases}$$

$$f_{X^2}(t) = \begin{cases} (1/\sqrt{2\pi})t^{-(1/2)}e^{-(t/2)} & t > 0 \\ 0 & t < 0 \end{cases}$$

Similarly

$$f_{Y^2}(t) = \begin{cases} (1/\sqrt{2\pi})t^{-(1/2)}e^{-(t/2)} & t > 0 \\ 0 & t < 0 \end{cases}.$$

We find from Proposition 2 and its corollary that

$$f_{X^2+Y^2}(a) = \int_{-\infty}^{\infty} f_{X^2}(t)f_{Y^2}(a - t)\,dt$$

$$= \int_{-\infty}^{a} \frac{1}{\sqrt{2\pi}} t^{-(1/2)}e^{-(t/2)} \cdot \frac{1}{\sqrt{2\pi}} (a - t)^{-(1/2)}e^{-((a-t)/2)}\,dt$$

for $a > 0$

$$= \frac{1}{2\pi} \int_{0}^{a} \frac{1}{\sqrt{t}\sqrt{a - t}} e^{-(t/2)}e^{-(a/2+t/2)}\,dt$$

$$= \frac{e^{-(a/2)}}{2\pi} \int_{0}^{a} \frac{1}{\sqrt{at - t^2}}\,dt$$

$$= \frac{e^{-(a/2)}}{2\pi} \left(\cos^{-1} \frac{a/2 - t}{a/2}\Big|_{0}^{a}\right)$$

$$= (e^{-(a/2)}/2\pi)\pi = (1/2)e^{-(a/2)}.$$

Thus

$$f_{X^2+Y^2}(a) = \begin{cases} (1/2)e^{-(a/2)} & \text{if } a > 0 \\ 0 & \text{if } a \le 0 \end{cases}.$$

(This density is known as a "chi-squared density with two degrees of freedom.")

We shall say that X and Y are "of the same type" provided they are either jointly discrete or jointly absolutely continuous.

Proposition 3: *Let X and Y be random variables of the same type and let $E(X)$ and $E(Y)$ exist. Then $E(X + Y)$ exists and $E(X + Y) = E(X) + E(Y)$.*

Proof: We shall treat only the jointly absolutely continuous case since the jointly discrete case is similar. Let $f_{X,Y}$ be a joint density for X and Y; then from Proposition 2, if $Z = X + Y$, we have that

$$f_Z(z) = \int_{-\infty}^{\infty} f_{X,Y}(x, a - x)\,dx;$$

so

$$E(Z) = \int_{-\infty}^{\infty} z f_Z(z)\, dz = \int_{-\infty}^{\infty} \int_{-\infty}^{\infty} z f_{X,Y}(x, z - x)\, dx\, dz$$

$$= \int_{-\infty}^{\infty} \int_{-\infty}^{\infty} (y + x) f_{X,Y}(x, y)\, dy\, dx \quad \text{where } y = z - x$$

$$= \int_{-\infty}^{\infty} \int_{-\infty}^{\infty} y f_{X,Y}(x, y)\, dy\, dx + \int_{-\infty}^{\infty} \int_{-\infty}^{\infty} x f_{X,Y}(x, y)\, dy\, dx$$

$$= \int_{-\infty}^{\infty} y \int_{-\infty}^{\infty} f_{X,Y}(x, y)\, dx\, dy + \int_{-\infty}^{\infty} x \int_{-\infty}^{\infty} f_{X,Y}(x, y)\, dy\, dx$$

$$= \int_{-\infty}^{\infty} y f_Y(y)\, dy + \int_{-\infty}^{\infty} x f_X(x)\, dx \text{ (Theorem 3 of section 7.2)}$$

$$= E(Y) + E(X).$$

Now the fact that $\int_{-\infty}^{\infty} |z| f_Z(z) f_Z(z)\, dz < \infty$ follows in a similar manner so $E(Z)$ *is* defined. $\square$

Example: A fair die is tossed and a die weighted such that the probability of either 1, 2, 3, 4, 5, or 6 is proportional to that number is tossed. Find the expected sum of the numbers obtained.

Let X be the r.v. giving the outcome of the first die and Y be that for the second die. Then we seek $E(X + Y) = E(X) + E(Y)$.

$$p_X(t) = \begin{cases} 1/6 & \text{if } t = 1, 2, 3, 4, 5, 6 \\ 0 & \text{if not} \end{cases}$$

so

$$E(X) = \sum_{i=1}^{6} \frac{1}{6}(i) = \frac{1}{6}(1 + 2 + 3 + 4 + 5 + 6) = \frac{21}{6} = 3.5.$$

$$p_Y(t) = \begin{cases} 1/21 & \text{if } t = 1 \\ 2/21 & \text{if } t = 2 \\ 3/21 & \text{if } t = 3 \\ \vdots & \\ 6/21 & \text{if } t = 6 \\ 0 & \text{otherwise} \end{cases} \quad \begin{array}{l} \text{(See the second example} \\ \text{of section 6.2.)} \end{array}$$

Thus

$$E(Y) = \sum_{i=1}^{6} \frac{i}{21} \, i = \frac{1}{21} \sum_{1}^{6} i^2$$

$$= (1/21)(1 + 4 + 9 + 16 + 25 + 36) = 91/21 = 4.333 \cdots;$$

so $E(X + Y) = 3.5 + 4.333 \cdots = 7.833 \cdots.$

Example: When two resistors are connected in series then the effective resistance is the sum of the individual resistances. Suppose one resistor has resistance determined by a r.v. X which is uniformly distributed between 700 and 900 (ohms); and another resistor has resistance determined by a r.v. Y whose density has a graph as given in figure 7.15.

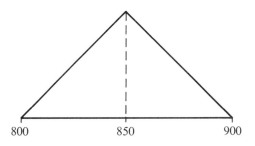

800   850   900

Figure 7.15

Compute the expected resistance of the paired connected in series. The resistance of the pair is given by a r.v. $Z = X + Y$.

$$E(X + Y) = E(X) + E(Y) = 800 + 850 = 1650.$$

Note, in the preceding result, that there is no assumption of independence. Note also that $E(X + Y) = \int_{-\infty}^{\infty} \int_{-\infty}^{\infty} (x + y) f_{X,Y}(x, y) \, dy \, dx$ or $E(X + Y) = \sum_{(x,y) \in R^2} (x + y) p_{X,Y}(x, y)$; the double integral and double sum can be considered a type of expectation with respect to both X and Y. This is similar to the situation of a single random variable when we investigated $E_X(g)$ for $g \colon R \to R$. If g is a Borel function then $E(g \circ X) = E_X(g)$. The same type of result holds for jointly distributed random variables. We prove this for the jointly discrete case, but the jointly absolutely continuous case requires much machinery which we have not developed.

THEOREM 1: Let X and Y be jointly distributed random variables and $g \colon R^2 \to R$ be a Borel function. Let $Z = g(X(s), Y(s))$; then Z is a random variable. Moreover,

(a)   If X and Y are jointly absolutely continuous, then

$$E(Z) = \int_{-\infty}^{\infty} \int_{-\infty}^{\infty} g(x, y) f_{X,Y}(x, y) \, dy \, dx$$

provided

$$\int_{-\infty}^{\infty} \int_{-\infty}^{\infty} |g(x, y)| f_{X,Y}(x, y) \, dy \, dx < \infty.$$

(b)   If X and Y are jointly discrete, then

$$E(Z) = \sum_{(x,y) \in R^2} g(x, y) p_{X,Y}(x, y)$$

provided

$$\sum_{(x,y) \in R^2} |g(x, y)| p_{X,Y}(x, y) < \infty.$$

Proof of (b):   The fact that Z is a random variable has already been established.

$$P(Z = z) = P(g(X, Y) = z) = P((X, Y) \in g^{-1}(z))$$

$$= \sum_{(x,y) \in g^{-1}(z)} p_{X,Y}(x, y) = \sum_{g(x,y)=z} p_{X,Y}(x, y).$$

$$E(Z) = \sum_{z \in R} z p_Z(z) = \sum_{z \in R} \left(\sum_{g(x,y)=z} z p_{X,Y}(x, y) \right)$$

$$= \sum_{z \in R} \left(\sum_{g(x,y)=z} g(x, y) p_{X,Y}(x, y) \right)$$

$$= \sum_{(x,y) \in R^2} g(x, y) p_{X,Y}(x, y) \quad \text{since as } z \text{ ranges over all R,}$$

then (x, y) ranges over all R^2 and so the double sum is simply a rearrangement of the single sum. $\square$

Example: Let X and Y be jointly absolutely continuous with joint density

$$f_{X,Y}(x, y) = \begin{cases} (x + y) & \text{if } (x, y) \in [0, 1] \times [0, 1] \\ 0 & \text{if not} \end{cases}.$$

Compute $E(X^2 + Y^2)$.

Here $g(x, y) = x^2 + y^2$ and by the preceding theorem,

$$E(X^2 + Y^2) = \int_{-\infty}^{\infty} \int_{-\infty}^{\infty} (x^2 + y^2) f_{X,Y}(x, y) \, dy \, dx$$

$$= \int_0^1 \int_0^1 (x^2 + y^2)(x + y) \, dy \, dx$$

$$= \int_0^1 \left(\int_0^1 x^3 + x^2 y + xy^2 + y^3 \, dy \right) dx$$

$$= \int_0^1 \left(x^3 + \frac{x^2}{2} + \frac{x}{3} + \frac{1}{4} \right) dx$$

$$= 1/4 + 1/6 + 1/6 + 1/4 = 10/12 = 5/6$$

Example: Two numbers are chosen at random from $[0, 1]$ on the basis of the uniform density. Find the expectation of the larger number.

Let X be the random variable determining one number and Y determine the other. We shall assume that

$$f_X(x) = f_Y(y) = \begin{cases} 1 & \text{on } [0, 1] \\ 0 & \text{otherwise} \end{cases}$$

and that X and Y are independent. Then

$$f_{X,Y}(x, y) = \begin{cases} 1 & \text{if } (x, y) \in [0, 1] \times [0, 1] \\ 0 & \text{if not} \end{cases}.$$

Let $g(x, y) = \max (x, y)$; then g is a Borel function from R^2 to R. We compute

$$E(\max (X, Y)) = \int_0^1 \int_0^1 \max (x, y) \cdot 1 \, dy \, dx$$

$$= \int_0^1 \left(\int_0^x \max (x, y) \, dy + \int_x^1 \max (x, y) \, dy \right) dx$$

$$= \int_0^1 \left(\int_0^x x \, dy + \int_x^1 y \, dy \right) dx$$

$$= \int_0^1 \left(x^2 + \frac{1}{2} - \frac{x^2}{2} \right) dx$$

$$= \frac{1}{2} \int_0^1 x^2 + 1 \, dx$$

$$= 2/3.$$

Example: Suppose that the resistances of two resistors are random variables X and Y related such that

$$f_{X,Y}(x, y) = \begin{cases} A(x + y) & \text{if } 50 \leq x, y \leq 100 \\ 0 & \text{if not} \end{cases}.$$

Suppose that the resistors are joined in parallel. Find the expected resistance of the system.

Recall that if resistors have resistance R_1 and R_2, then the effective resistance for the two in parallel is R where

$$1/R = 1/R_1 + 1/R_2 \quad (\text{i.e. } R = (R_1 \cdot R_2)/(R_1 + R_2)).$$

Let $Z = (X \cdot Y)/(X + Y)$; then we must find $E(Z)$. By computing $\int_{50}^{100} \int_{50}^{100} x + y \, dy \, dx$ we find that $A = 1/375{,}000$ and so by the preceding theorem,

$$E(Z) = \frac{1}{375{,}000} \int_{50}^{100} \int_{50}^{100} \left(\frac{xy}{x + y}\right)(x + y) \, dx \, dy$$

$$= \frac{1}{375{,}000} \int_{50}^{100} \int_{50}^{100} xy \, dy \, dx = \frac{1}{375{,}000} \left.\frac{x^2}{2}\right|_{50}^{100} \cdot \left.\frac{y^2}{2}\right|_{50}^{100}$$

$$= (1/50 \cdot 7500)(7500/2)^2 = 7500/200 = 37.5.$$

Definition 1: Let X and Y be random variables of the same type and let $\mu_X = E(X)$ and $\mu_Y = E(Y)$. The *covariance of X and Y* is defined to be

$$\text{Cov}\,(X, Y) = E((X - \mu_X) \cdot (Y - \mu_Y))$$

if this expectation exists.

We shall consider certain purely algebraic properties of the covariance before we consider its use in probability.

THEOREM 2: Let X, Y, and Z be random variables of the same type. Then (assuming appropriate existence)

(a) $\text{Cov}\,(X, X) = \text{Var}\,(X)$,

(b) $\text{Cov}\,(X, Y) = \text{Cov}\,(Y, X)$,

(c) $\text{Cov}\,(X, Y) = E(X \cdot Y) - E(X) \cdot E(Y)$,

(d) Cov $(kX, Y) = k$ Cov (X, Y) for each $k \in$ R

(e) Cov $(X, kY) = k$ Cov (X, Y) for each $k \in$ R,

(f) Cov $(X, Y + Z) =$ Cov $(X, Y) +$ Cov (X, Z),

(g) Cov $(X + Y, Z) =$ Cov $(X, Z) +$ Cov (Y, Z).

Proof: Parts (a) and (b) are obvious; (e) follows from (d) and (b); (g) follows from (f) and (b).

(c) $$\begin{aligned} \text{Cov } (X, Y) &= E((X - \mu_X) \cdot (Y - \mu_Y)) \\ &= E(X \cdot Y - \mu_X Y - \mu_Y X + \mu_X \mu_Y) \\ &= E(X \cdot Y) - \mu_X E(Y) - \mu_Y E(X) + \mu_X \mu_Y \\ &= E(X \cdot Y) - \mu_X \mu_Y - \mu_X \mu_Y + \mu_X \mu_Y \\ &= E(X \cdot Y) - E(X) \cdot E(Y). \end{aligned}$$

(d) $$\begin{aligned} \text{Cov } (kX, Y) &= E(kX \cdot Y) - E(kX) \cdot E(Y) \\ &= k(E(X \cdot Y) - E(X) \cdot E(Y)) \\ &= k \text{ Cov } (X, Y). \end{aligned}$$

(f) $$\begin{aligned} \text{Cov } (X, Y + Z) &= E(X \cdot (Y + Z)) - E(X) \cdot E(Y + Z) \\ &= E(X \cdot Y + X \cdot Z) - E(X)[E(Y) + E(Z)] \\ &= E(X \cdot Y) + E(X \cdot Z) - E(X) \cdot E(Y) \\ &\quad - E(X) \cdot E(Z) \\ &= \text{Cov } (X, Y) + \text{Cov } (X, Z). \quad \square \end{aligned}$$

Example: Find the covariance of the random variables Z and W whose joint density is given by

$$f_{Z,W}(s, t) = \begin{cases} s + t & \text{if } 0 \le s, t \le 1 \\ 0 & \text{if not} \end{cases}$$

We have treated this joint density before (section 7.2) and found that

$$f_Z(s) = \begin{cases} s + 1/2 & \text{if } 0 \le s \le 1 \\ 0 & \text{if not} \end{cases}$$

Thus

$$E(Z) = \int_0^1 s^2 + s/2 \, ds = 1/3 + 1/4 = 7/12$$

and similarly, $E(W) = 7/12$.

$$E(Z \cdot W) = \int_0^1 \int_0^1 (st)(s + t) \, ds \, dt$$

$$= \int_0^1 \int_0^1 s^2 t + t^2 s \, ds \, dt$$

$$= 1/3$$

and so

$$\text{Cov } (Z \cdot W) = E(Z \cdot W) - E(Z) \cdot E(W)$$

$$= 1/3 - (7/12)^2$$

$$= -(1/144).$$

Example: Find the covariance of the random variables X and Y in the last resistor example.

In this case the joint density for X and Y is given by

$$f_{X,Y}(x, y) = \begin{cases} (1/375,000)(x + y) & \text{if } 50 \le x, y \le 100 \\ 0 & \text{if not} \end{cases}.$$

It is not hard to show that if Z and W are as in the preceding example, then $X = 50Z + 50$ and $Y = 50W + 50$. Thus

$$\text{Cov } (X, Y) = \text{Cov } (50Z + 50, 50W + 50)$$

$$= (50)^2 \text{ Cov } (Z + 1, W + 1)$$

$$= (2500)(E((Z + 1) \cdot (W + 1)) - E(Z + 1) \cdot E(W + 1))$$

$$= (2500)(E(ZW + Z + W + 1) - (E(Z) + 1)(E(W) + 1))$$

$$= (2500)(E(ZW) + E(Z) + E(W) + 1 - (E(Z)E(W) + E(Z) + E(W) + 1))$$

$$= 2500(E(ZW) - E(Z)E(W))$$

$$= 2500 \text{ Cov } (Z, W)$$

$$= 2500(-(1/144))$$

$$= -2500/144.$$

As the name indicates, the covariance is a measure of how the random variables X and Y vary together—a measure of how they can be related. If X and Y are independent, then in a sense, they are not related at all (knowing the value of one tells nothing about the value of the other); in this case the covariance is 0.

Proposition 4:   Let X and Y be independent random variables of the same type. If Cov (X, Y) exists, then Cov $(X, Y) = 0$.

Proof: We shall prove the theorem for the absolutely continuous case since the proof for the discrete case is similar.

$$\text{Cov } (X, Y) = E(X \cdot Y) - E(X) \cdot E(Y)$$

$$= \int_{-\infty}^{\infty} \int_{-\infty}^{\infty} xy \, f_{X,Y}(x, y) \, dy \, dx - E(X)E(Y)$$

$$= \int_{-\infty}^{\infty} \int_{-\infty}^{\infty} xy \, f_X(x) f_Y(y) \, dy \, dx - E(X)E(Y)$$

$$= \left(\int_{-\infty}^{\infty} x f_X(x) \, dx \right) \left(\int_{-\infty}^{\infty} y f_Y(y) \, dy \right) - E(X)E(Y)$$

$$= 0. \quad \square$$

Unfortunately, the converse of the preceding proposition is not true; there do exist random variables X and Y such that Cov $(X, Y) = 0$ but X and Y are not independent.

The following two results indicate ways in which the covariance of X and Y is related to the individual variances of X and Y.

Proposition 5:   If X and Y are of the same type, then

$$\text{Var } (X + Y) = \text{Var } (X) + \text{Var } (Y) + 2 \text{ Cov } (X, Y);$$

in the sense that if either side exists, then both exist and are equal.

Proof:

$$\text{Var } (X + Y) = E((X + Y)^2) - [E(X + Y)]^2$$

$$= E(X^2 + 2XY + Y^2) - (E(X) + E(Y))^2$$

$$= E(X^2) + 2E(XY) + E(Y)^2 - (E(X))^2$$

$$\quad - 2E(X)E(Y) - (E(Y))^2$$

$$= \text{Var } (X) + \text{Var } (Y) + 2 \text{ Cov } (X, Y). \quad \square$$

Proposition 6:   (*Cauchy–Schwarz Inequality*)   *Let X and Y be random variables of the same type. Then*

$$(\text{Cov } (X,\, Y))^2 \le \text{Var } (X) \cdot \text{Var } (Y)$$

(if they exist), with equality only when there exist constants a, b, and c such that $P(aX - bY = c) = 1$ and either $a \ne 0$ or $b \ne 0$.

Proof: If Cov $(Y,\, Y) = 0$, then Cov $(X,\, Y) = 0$ (why?) and the proposition is proved.   Otherwise, let a and b be real numbers and $W = aX - bY$; then

$$0 \le \text{Var } (W) = \text{Cov } (aX - bY,\, aX - bY)$$
$$= a^2 \text{ Cov } (X,\, X) - 2ab \text{ Cov } (X,\, Y) + b^2 \text{ Cov } (Y,\, Y),$$

with equality only when there is a constant c such that $P(W = c) = 1$ (see exercise 8, section 6.3).

In particular, if $a = \text{Cov } (Y,\, Y)$ and $b = \text{Cov } (X,\, Y)$, then we have that

$$0 \le [\text{Cov } (Y,\, Y)]^2 \text{ Cov } (X,\, X) - 2 \text{ Cov } (Y,\, Y)[\text{Cov } (X,\, Y)]^2$$
$$+ [\text{Cov } (X,\, Y))^2 \text{ Cov } (Y,\, Y)$$
$$= \text{Cov } (Y,\, Y) (\text{Cov } (Y,\, Y) \text{ Cov } (X,\, X) - [\text{Cov } (X,\, Y)]^2)$$
$$= \text{Var } (Y)(\text{Var } (Y) \text{ Var } (X) - [\text{Cov } (X,\, Y)]^2).$$

Thus Var (Y) Var $(X) - [\text{Cov } (X,\, Y)]^2 \ge 0$ from which the Proposition follows. □

The preceding proposition implies that

$$\frac{[\text{Cov } (X,\, Y)]^2}{\text{Var } (X) \text{ Var } (Y)} \le 1;$$

we define the *correlation coefficient of X and Y* by

$$\rho_{X,Y} = \frac{\text{Cov } (X,\, Y)}{\sqrt{\text{Var } (X) \text{ Var } (Y)}}.$$

It is clear that

(1) $-1 \le \rho_{X,Y} \le 1$

(2) $\rho_{X,Y} = \pm 1$ only if there exist constants a, b, and c such that $P(aX - bY = c) = 1$,

(3)   if X and Y are independent, then $\rho_{X,Y} = 0$.

$\rho_{X,Y}$ is a measure of the accuracy of a linear prediction between X and Y since when $\rho_{X,Y} = \pm 1$, then one may predict X from Y or Y from X with probability 1; but when $\rho_{X,Y} = 0$, then no linear prediction can be made with a high degree of accuracy. If $\rho_{X,Y} = 0$, we may say that X and Y are *uncorrelated*.

Example: Find the correlation coefficient for the random variables X and Y in the preceding example.

Again we introduce Z and W such that $X = 50Z + 50$ and $Y = 50W + 50$ and we compute

$$E(Z) = \int_0^1 s\left(s + \frac{1}{2}\right) ds = \frac{7}{12} = E(W)$$

$$E(Z^2) = \int_0^1 s^2\left(s + \frac{1}{2}\right) ds = \frac{10}{12} = E(W^2)$$

$$E(ZW) = \int_0^1 \int_0^1 st(s + t)\, ds\, dt = \frac{1}{3}.$$

Thus Var (Z) = Var (W) = $10/12 - (7/12)^2 = 71/144$ and Cov $(Z, W) = 1/3 - (7/12)^2 = -1/144$ and so

$$\rho_{Z,W} = \frac{-(1/144)}{\sqrt{(71/144) \cdot (71/144)}} = -\frac{1}{71}.$$

Now by exercise 12 of this section, $\rho_{X,Y} = -(1/71)$ also.

If a person tosses a fair coin ten times and obtains a head on each toss, then he often feels that a tail will be more likely on the 11th toss because things have to "average out" in the end. In a sense, this "averaging out" usually does take place—but only in the limit (if $\{a_n\}_1^\infty$ is a sequence of positive numbers and $\lim_{n \to \infty} a_n = 0$, then we know that eventually, $a_n < 1/100$ but we may have to consider several million terms before we can ensure that terms be this small). The following result is often misquoted as the "Law of Averages."

WEAK LAW OF LARGE NUMBERS: Let $\{X_1, X_2, \ldots\}$ be a pairwise uncorrelated sequence of random variables (if $i \neq j$, then $\rho_{X_i, X_j} = 0$) all of the same type. Let $\mu \in \mathbb{R}$ and $\sigma > 0$ such that for each $n \in \mathbb{N}$, $E(X_n) = \mu$ and Var $(X_n) = \sigma^2$. For each $n \in \mathbb{N}$, let $S_n = 1/n \sum_{i=1}^{n} X_i$; then for each $\varepsilon > 0$, $\lim_{n \to \infty} P(|S_n - \mu| \leq \varepsilon) = 1$.

Proof: First of all, note that S_n is a random variable and

$$E(S_n) = E\left(\frac{1}{n}\sum_1^n X_i\right) = \frac{1}{n}\sum_1^n E(X_i) \quad \text{by Proposition 3,}$$

$$= (1/n)(n\mu)$$

$$= \mu.$$

Since the X_n's are pairwise uncorrelated, then for $i \neq j$, we have Cov $(X_i, X_j) = 0$ and thus by Proposition 5, Var $(X_i + X_j) =$ Var (X_i) + Var (X_j). A simple induction argument establishes a similar formula for the sum of any finite number of the X_n's. Thus

$$\text{Var } (S_n) = \text{Var}\left(\frac{1}{n}\sum_1^n X_i\right) = \frac{1}{n^2}\text{Var}\left(\sum_1^n X_i\right)$$

$$= \frac{1}{n^2}\sum_1^n \text{Var } (X_i)$$

$$= (1/n^2)(n\sigma^2)$$

$$= \sigma^2/n.$$

We now use Chebyshev's inequality for S_n and find that for $h = \sqrt{n}\varepsilon/\sigma$ (i.e. $\varepsilon = h(\sigma/\sqrt{n})$) we have

$$P(|S_n - \mu| \leq \varepsilon) = P(\mu - \varepsilon \leq S_n \leq \mu + \varepsilon)$$

$$\geq P(\mu - \varepsilon < S_n \leq \mu + \varepsilon)$$

$$\geq 1 - \frac{1}{h^2} = 1 - \frac{1}{n\varepsilon^2/\sigma^2}$$

$$= 1 - \sigma^2/(n\varepsilon^2) \quad \text{which has limit 1 as in } n \to \infty. \quad \square$$

Another form of the conclusion is that $\lim_{n\to\infty} P(|S_n - \mu| > \varepsilon) = 0$. The Weak Law of Large Numbers has the following interpretation in terms of independent trials: if X_1 is the outcome on the first trial, X_2 is the outcome on the second trial, etc.; then the probability that the average of the outcomes differs from the "theoretical" expectation by more than some fixed amount goes to 0 as the number of trials increases.

When we introduced the concept of expectation we looked at an experiment which has as outcomes the numbers $\{a_1, \ldots, a_n\}$ with probabilities

$p_1, p_2, \ldots, p_n$ respectively. For a large number (N) of repetitions of the experiment, we "expected" a_i to occur approximately $p_i \cdot N$ times. In a sense, this expectation is borne out by the following corollary.

Corollary: Let A be an event in a particular experiment and $P(A) = p$. Suppose that the experiment is repeated independently N times and k_N is the number of times that A occurs. Then for each $\varepsilon > 0$,

$$\lim_{N \to \infty} P\left(\left|\frac{k_N}{N} - p\right| \leq \varepsilon\right) = 1.$$

Proof: For each $m \in \mathbf{N}$, let

$$X_m = \begin{cases} 1 & \text{if } A \text{ occurs on the } m\text{th trial} \\ 0 & \text{if not} \end{cases}.$$

We assume that the X_m's are independent so thus they are uncorrelated. For each m,

$$E(X_m) = \sum_{x \in \mathbf{R}} x P(X_m = x) = 0 \cdot P(X_m = 0) + 1 \cdot P(X_m = 1) = p.$$

It is easy to see that the variances are all equal since the X_m's share the same probability mass function. Now with the notation of the Weak Law of Large Numbers,

$$S_N = \frac{1}{N} \sum_{i=1}^{N} X_i = \frac{1}{N} (k_N)$$

$$= k_N/N.$$

Thus

$$1 = \lim_{N \to \infty} P(|S_N - p| \leq \varepsilon) = \lim_{N \to \infty} P\left(\left|\frac{k_N}{N} - p\right| \leq \varepsilon\right).$$

Example: Experimentally find the probability that a thumbtack, when tossed, will land with its point upward.

Obviously, we cannot find the exact probability, or even be absolutely sure that we are within, say, 0.01 of the exact probability regardless of how many times the thumbtack is tossed. But we can use the Weak Law of Large Numbers to find how many times to toss the tack so that the probability is 0.90 that the percentage of times that the tack points upward is within 0.01 of the true probability.

As in the proof of the corollary, let

$$X_m = \begin{cases} 1 & \text{if the tack points upward on the } m\text{th toss .} \\ 0 & \text{if not} \end{cases}$$

Then $E(X_m) = P(\text{tack points upward}) = p$, and

$$\text{Var } (X_m) = E(X_m^2) - E(X_m)^2$$

$$= (1)^2 p + (0)^2(1 - p) - (p^2)$$

$$= p - p^2 = p(1 - p).$$

If

$$S_N = \frac{1}{N} \sum_{i=1}^{N} X_i,$$

then

$$E(S_N) = k_N/N$$

as before and

$$\text{Var } (S_N) = \frac{1}{N^2} \sum_{1}^{N} p(1 - p) = \frac{p(1 - p)}{N} .$$

Therefore

$$P(|S_N - p| \le \varepsilon) \ge 1 - p(1 - p)/N\varepsilon^2.$$

So to ensure that $P(|S_N - p| \le 0.01) \ge 0.9$, we need only require that

$$1 - (p(1 - p)/N(0.01)^2) \ge 0.9$$

or

$$p(1 - p)/N(0.01)^2 \le 0.1$$

which is equivalent to

$$N \ge p(1 - p)/(0.1)(0.01)^2 = p(1 - p)/0.0001.$$

This is an unsatisfactory answer since the whole problem was to approximate p. Now $y = x(1 - x)$ is the equation of a parabola which reaches its maximum at $x = 1/2$ at which $y = 1/4$. Thus $p(1 - p) \le 1/4$ for each p. So if $N \ge (1/4)(1/0.0001)$ then $N \ge p(1 - p)/0.0001$; thus $N \ge 2500$.

Exercises:

1. Find the probability mass function for $Z = X + Y$ when

$$p_{X,Y}(x, y) = \begin{cases} 1/6 & \text{at } (0, 1), (1, 0), (0, 0) \\ 1/2 & \text{at } (2, 2) \\ 0 & \text{otherwise} \end{cases}$$

2. Find a density for $Z = X + Y$ when

$$f_{X,Y}(x, y) = \begin{cases} 4xy & \text{if } (x, y) \in [0, 1] \times [0, 1] \\ 0 & \text{if not} \end{cases}.$$

3. Find a density for $Z = X^2 + Y^2$ where

$$f_{X,Y}(x, y) = \begin{cases} 1 & \text{if } (x, y) \in [0, 1] \times [0, 1] \\ 0 & \text{if not} \end{cases}.$$

4. Find $E(X^2 + Y)$ where X and Y are as in 1.

5. Find $E(X^2 + Y)$ where X and Y are as in 2.

6. Prove Proposition 3 in the jointly discrete case.

7. Find Cov (X, Y) for X and Y as in 1.

8. Find Cov (X, Y) for X and Y as in 2.

9. Show that if X and Y are uncorrelated and X and Z are uncorrelated, then X and $Y + Z$ are uncorrelated.

10. Prove Proposition 4 in the jointly discrete case.

11. Use the Cauchy–Schwartz inequality to show that if X and Y are random variables of the same type then $\sigma_{X+Y} \leq \sigma_X + \sigma_Y$ (if they exist) (recall $\sigma_X = \sqrt{\text{Var}(X)}$).

12. Suppose that X and Y are random variables and that $a, b, \in \text{R}$. Show that Cov $(aX + b, Y) = a \cdot \text{Cov}(X, Y)$. Show that $\rho_{aX+b,Y} = \rho_{X,Y}$.

13. Two independent resistors have as resistance random variables X and Y uniformly distributed on $[50, 100]$. The resistors are wired up in parallel and Z is the effective resistance of the system. Set up the integrals required to find $E(Z)$ (the *really* ambitious student should attempt to evaluate it).

Index

Absolute continuity, 86
  density, 86
  expectation, 108
  joint absolute continuity, 145
  sum of jointly absolutely continuous
    r.v.'s, 159
Admissible set, 144
Algebra, 12

Bayes' Theorem, 52
Bernoulli trials, 64
Binomial probability mass function, 98
  moments, 123
  moment generating function, 123
Binomial Theorem, 41
Borel function, 102
Borel sets:
  in R, 28
  in R^n, 102
Boundary, 143
Buffon's Needle Problem, 157

Cauchy density, 98
  lack of expectation, 109
Cauchy-Schwarz inequality, 176
Cell, 41
Central moments, 126
Characteristic function, 74
Chebyshev's Inequality, 127
Chi-Squared distribution, 166, 167
Closed set, 105
Co-finite, 12
Combination, 39
Complement, 4, 5
  relative complement, 4
Conditional probability, 48
Continuity:
  absolute continuity, 86

Continuity (*continued*)
  continuity from above, 19
  continuity from below, 19
  continuity from the right, 29
  continuity in R, 101
  piecewise continuity on R, 86
  piecewise continuity on R^2, 145
Content 0, 142
Convex combination, 21
Convolution, 160
Correlation coefficient, 176
Countable, 11
Countably additive, 17
Covariance, 172
  algebraic properties, 172, 175, 176

Decomposable r.v., 95
  expectation, 110
DeMorgan's laws, 5
  extended, 10
Density, 86
  joint, 146
Discrete probability measure space, 23
Discrete r.v., 82
  expectation, 107
  jointly discrete, 138
  probability mass function, 82
  sum of jointly discrete r.v.'s, 159
Distribution function:
  joint, 131
  marginal, 136
  for probability measure on $\mathscr{B}(R)$, 29
  for a random variable, 75, 77, 78
Distributive Laws, 4
  extended, 10
Dominated Convergence Theorem, 122

Event, 17
  independent events, 56, 60, 61
Equiprobable measure, 22
Expectation:
  absolutely continuous r.v., 108
  discrete r.v., 107
  decomposable r.v., 110
  of a function with respect to a r.v.,
    113
  of a function of a r.v., 116, 118
  of a function of two r.v.'s, 170
  of a sum of r.v.'s, 167
Exponential distribution, 90
  density, 98
  expectation, 112

Factorial, 35
  Stirling's approximation, 65
Finite additivity, 17
Fundamental Principle of Counting, 34

Independent events, 56, 60, 61
Independent random variables, 151,
  157
  sum of absolutely continuous inde-
    pendent r.v.'s, 160
  sum of discrete independent r.v.'s,
    160
  product of independent uniform
    r.v.'s, 164
  sum of independent uniform r.v.'s,
    162
  sum of independent Poisson r.v.'s,
    161
  joint density for absolutely con-
    tinuous r.v.'s, 152
  joint probability mass function for
    discrete r.v.'s, 151
Indexed class, 8
Indicator function. See Characteristic
  function
Intersection, 3, 8

Joint distribution, 131
Jointly absolutely continuous r.v.'s,
  145

Jointly absolutely continuous r.v.'s
  (continued)
  joint density, 146
  joint density for independent r.v.'s,
    152
Jointly discrete r.v.'s, 138
  joint probability mass function, 138
  joint probability mass function for
    independent r.v.'s, 151

Lattice, 6

Marginal distribution, 136
Mean. See Expectation
Mean deviation, 126
Modular, 17
Moments of a r.v., 120
  central moments, 126
  moment generating function, 122
Monotone, 17

Normal distribution, 90
  normal density, 90, 105
  expectation, 109

Open set, 100
Ordered partition, 41

Partition, 41
  ordered partition, 41
Permutation, 36
Piecewise continuity on R, 86
Piecewise continuity on R^2, 145
Poisson measure, 67
Poisson probability mass function, 98
  sum of independent Poisson r.v.'s,
    161
Probability mass function, 82
  joint probability mass function, 138
Probability measure space, 17
  as model, 20

Random variable (r.v.), 70
  absolutely continuous r.v., 86
  decomposable r.v., 95
  discrete r.v., 82

Relative complement, 4
Repeated trials, 64
Ring, 7

Sample space, 17
Set, 1
Set function, 16
Sigma algebra, 12
Sigma ring, 11
Standard deviation, 126
Standardization, 127
Stirling's approximation, 65
Subtractive, 17
Symmetric difference, 6

Tree diagrams, 49

Two additive, 16

Uniform distribution, 79
    density, 98
    expectation, 108
    product of independent uniform
        r.v.'s, 164
    sum of independent uniform r.v.'s,
        162
Union, 3, 8

Variance, 126

Weak law of large numbers, 177
Weight function, 23